Street Smart Firefighting !

The Common Sense Guide To Firefighter Safety and Survival

Robert C. Bingham

Street Smart Firefighting!

The Common Sense Guide To Firefighter Safety and Survival

Robert C. Bingham

Valley Press
Post Office Box 2044
Vienna, Virginia 22183
1-888-825-9550
streetsmartfire.com

Printed in the United States of America

Library of Congress Cataloging-in-Publication Data
Bingham, Robert C.
Street Smart Firefighting / Robert C. Bingham
          p.   cm.
LCCN 2003117034
ISBN 0-9748447-0-5 (alk. Paper)
1. Fire Extinction- suppression
TH9145. B56 2004
628.9'25-dc20

Notice to the reader

The reader is expressly warned to consider and adopt all safety precautions that might be indicated by the information provided and to avoid all potential hazards. By following the instructions in this book, the reader willingly assumes all risks in connection with such instruction.

DEDICATION

This book is dedicated to all the firefighters in America who are on the front lines every day.

CHAPTERS

FOREWORD

My fire service career was like no other. Years ago, I was an executive and a very enthusiastic New York Fire Buff, intensely interested in the art of fire suppression. I spent many hours in New York and some in Boston, riding and listening to experienced fire officers and picking their brains. I spent many hours in the New York Public Library studying back issues of *Fire Engineering*. I gained valuable but limited hands on experience as a member of the pre-war FDNY Auxiliary Corps. I enlisted in the Navy in January 1942. I was assigned to Panama, became a commissioned officer and was assigned to the Navy fire fighting program. The officers for this program were individually recruited from the ranks of large fire departments and under the command of Navy Captain Harold Burke, on military leave from his post as a Deputy Chief of FDNY. (Later Chief of Department.)   I was able to work side by side with these officers because of the ability I had gained by self-study and the great mentoring that I had received. For instance, I was able to whip a puzzling ship fire by applying a lesson I had learned in a discussion of another fire that showed the importance of learning the layout and construction features. When the war ended I was asked by the Navy to take my same officer position as a senior civilian employee.

The point is that due to multiple shifts and the reduction of fires, many officers today really have not had very much heavy-duty fire experience before they are promoted. Good mentoring can help make up for the deficiency in experience, as it did for me.

Chief Bingham's book is a triumph of good mentoring, the sort I received verbally from an earlier generation. He has been there and done that. This book is the most empathetic, yet no-nonsense, practical and "tell-it-how-it-is" resource you will find. It is an immensely useful guide, filled with new ideas you can use, and practical advice from a very experienced and capable fire chief. The advice is sound, insightful, and can be used by rookie firefighters, experienced chiefs, and everybody in between. He understands how firefighters think, where they make their biggest mistakes, and what they need to know to work safer and smarter. Bob tells many true stories and speaks in a down-to-earth style, and that makes for a good easy read.

I have often thought that a top-notch fire officer should write down what he has learned through the years, to pass on to future generations of firefighters. Bob Bingham has done this in an interesting and instructive manner.

Bob drives his points home in a way that is not intimidating, and makes learning fun and interesting. The book is filled with common sense and introduces new concepts. When you are finished reading this book you will be better prepared to do your job, and it could save your life.

Francis L. Brannigan SFPE (fellow)
Author of *BUILDING CONSTRUCTION FOR THE FIRE SERVICE*

PREFACE

This book is aimed at all firefighters, career and volunteer, in departments ranging from the smallest to the largest. It is intended to be a "reality check" and a guideline to improve the operations of our fire departments. The book identifies the key elements of safe, successful and predictable fireground operations. It is filled with understandable and sometimes amusing short stories and examples. ***All the stories, people and examples are real.*** The names may be changed to protect the innocent (and the guilty) but all the stories are true. The stories come from fire departments around the country, including my personal experiences as a deputy chief in the District of Columbia Fire Department.

Why You Need This Book

With so many fire books on the market, why do you need this one? Because this is the only fire service book that contains just about everything that you need to know to survive on the fireground. It will guide you towards safe and efficient fireground and emergency operations using tried and true procedures and good old common sense.

This book is different from the other fire fighting guides because it introduces a new way to do business. Books currently available to us in the fire service are often heavy on theory and light on reality. The textbooks will tell you that there are fifteen size-up considerations or tell you to follow steps A, B, C and D to extinguish the fire. The information is technically correct, but often does not reflect the real world. The lessons in this book are based on real life experiences.

This book has a fresh approach and includes chapters on philosophy, dispatch procedures, and preparation. All these subjects greatly affect fireground performances, and are often neglected in the typical fireground books. For example, our philosophy is very important because it drives our goals and how we do business at an emergency scene. Often the success or failure of an operation depends upon how we think.

Most fire departments are under staffed, under trained and under funded. In many areas, medical calls are increasing, while fires are declining. The result is that many departments are having problems handling fire calls, and there is a great need for a practical approach to fire fighting. This book provides the basis for developing such an approach.

This book provides little information about selecting the right nozzle, raising ladders or operating pumps. Rather, this book emphasizes avoiding common pitfalls and using proven methods to improve operations. I have focused on areas that I believe need improvement in most fire departments. The aim is to address our most common problems; most of the principles covered in the book apply to all incidents.

Some of the information in this book may be unconventional because it challenges popular opinion, examines long held beliefs and generally kicks up a lot of dust. It's time to shake the tree of tradition to get people thinking about the best ways to operate. We cannot simply accept things because "that is the way we've always done them." The fire service has changed and we must adapt by using a practical approach to our work.

Why I Wrote This Book

I considered writing this book for several years. As a fire service instructor, around the country, I have taught many command courses that include videos of fireground "bloopers" showing safety and tactical errors captured on film during actual fireground operations. The reactions from the students range from amusement to disbelief. Students often want to know which fire departments are shown committing such obvious and serious errors. My answer is that it is all of our fire departments, yours and mine. I then ask if there are any people in the class who have never seen any of these problems occur on their own firegrounds. No student has ever raised a hand.

When I stand in front of a command class, I always ask the following questions.

- "Have any of you ever been at an emergency scene and wondered if there was a plan?"

- "Have you ever been embarrassed by some of the things you've seen?

- "Have you ever gone back to the station and said, we were just lucky this time?"

Invariably, all the faces would light up and there would be plenty of nods. I have connected with them through the reality that **many fire departments have fire and emergency operations that leave a lot to be desired.**

During my 31 years in the fire service, I have come across almost every type of situation imaginable, and a few that were unimaginable. I learned some lessons the hard way, sometimes by burning down buildings, sometimes by taking casualties that could have been prevented. I have witnessed many fire departments in action all over the country and observed what works, what doesn't work and why. Through experience, I have become street smart, which is the ability to know what is happening, and what is likely to happen. Although you can't learn "street smarts" from a book, you can learn to think smart, to challenge current practices, and operate more efficiently and safely.

Acknowledgments

A very special group helped as content reviewers for this book. They helped me to improve the text and I thank them for their efforts. In alphabetical order they are:

Raymond Askins
Shellsburg, Fire Department
Shellsburg, Pennsylvania

Daniel G. Bingham
Captain
Arlington County Fire Department
Arlington, Virginia

James Embrey
Lieutenant (retired)
District of Columbia Fire Department
Washington D.C.

Joseph Herr
Fire Chief
Howard County, Maryland

David L. Rohr
Deputy Chief
Fairfax County Fire Department
Fairfax, Virginia.

Chief J.Gordon Routley
Fire Service Consultant
Montreal, Canada

Special thanks to my son Danny Bingham and my editor Andrea Kuettel for all their help and advice.

Cover photo by Lieutenant William Rabbit, District of Columbia Fire Department.

Thanks to Scotty W. Boatright, photographer, Fairfax County Fire Department, Fairfax, Virginia.

Line drawn illustrations by John York and Mike McGurk

1

THE WELL-RUN FIREGROUND

Proper fireground operations begin at the fire station. Imagine the following structure fire where key fireground principles are applied. All equipment has been carefully checked and is in working order. At the Dispatch Center, the 9-1-1 line buzzes; the time is 04:01 hours. An excited caller reports smoke in a building on Main Street, an older part of town that has seen better days. The standard response, of two engines, a truck company, a chief and an ambulance, are dispatched. Before fire companies arrive, the Dispatch Center is flooded with calls, all reporting the Main Street fire. Several callers report people trapped on the upper floors of the building.

The fire department's philosophy is to increase response levels when there is a strong possibility of a working incident, particularly when life threats are reported. Therefore, a third engine, another truck company and a medic unit are automatically dispatched. The department has found that an automatic dispatch system speeds up their operations and improves firefighter safety.

First Engine

The first due engine arrives and reports:

Engine 1 laid out to a hydrant at 2nd and Main.

Engine 1 on the scene of a three-story, semi-detached commercial with apartments above, ordinary construction, heavy smoke showing from the first floor, passing Command.

This fire department requires the first arriving engine to establish its own water supply, by laying a supply line from a hydrant if possible, to ensure a reliable water supply. The engine assumes a position in front of the building, allowing room for the truck company. The first engine company's plan is simple: go inside the store, locate the seat of the fire and extinguish it. Although there is a report of people trapped above the fire, they will try for rapid extinguishment because putting out the fire usually eliminates rescue problems.

3

THE ENGINE ATTACKS

Command

Command is established quickly.

The truck arrives and the officer assumes command of the fireground from an outside position. After looking over the situation, the truck officer (as Incident Commander) reports a working fire. This results in an automatic working fire dispatch of an air supply truck, a truck company for rapid intervention, another chief for the safety officer position and a fire investigator. This dispatch sends the special equipment and people usually needed at a working fire.

Ladder Truck

They vent quickly, working in teams.

The truck company goes into action, working in teams, so that no member operates alone. Venting quickly, they knock out the windows using ladders and tools, and clear away any obstructions to the venting. Their philosophy is that glass is cheap and fire fighting inside poorly ventilated structures is dangerous. The aerial ladder is raised to the roof, the skylight is removed and roof conditions are reported to Command. The roof ventilation releases smoke and heat and improves visibility. This improves the chances of survival for both victims and firefighters. Proper ventilation also reduces or eliminates the chances of flashover or backdraft. After the roof team ventilates, they quickly get off the roof because the longer they are there, the greater the danger. In any case, firefighters are not permitted to work directly over fires on any roof.

THE FIREGROUND SITUATION

Search

They start searching for those in the greatest danger first.

The truck company begins to search for victims, using a two-member team for speed and safety. No one searches alone. They search first for those in the greatest danger. Since it is unlikely that the hardware store is occupied at this time of night, the truck company team starts searching the second floor apartments, the most likely place to find victims. The top floor apartments are searched next, followed by the first floor store. At the same time, a primary search is made by a second team, along with advancing hose lines on the first floor.

Command

The Incident Commander (IC) quickly develops a plan based on what they are trying to accomplish.

The chief arrives and parks the vehicle in the front, with a good view of the building. The Incident Commander consults with the truck officer, establishes a stationary command post and assumes command. The IC's size-up indicates that the fire is in a hardware store on the first floor of a three-story semi-detached building. There is a potential for loss of life and property on the floors above and in the exposure building.

The IC quickly develops a plan based on what they are trying to accomplish. The plan is: extinguish the fire, rescue people above the fire, establish medical treatment capability, and protect the exposures. This plan is simple, but covers the important concerns.

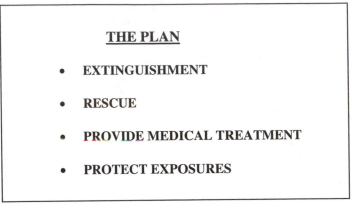

THE PLAN

- **EXTINGUISHMENT**

- **RESCUE**

- **PROVIDE MEDICAL TREATMENT**

- **PROTECT EXPOSURES**

Standard Operating Procedures (SOPs)

Using SOPs, fire operations follow a pattern, and the same things happen at each fire.

When the working fire is verified, all first alarm companies follow the departments standard operating procedures for structure fires without specific direction from Command. This is called management by exception, relieving the IC from having to tell everybody what to do at every incident. The IC then has time to develop plans and think ahead. The SOPs are simple, easily understood and cover most fire situations.

Using SOPs, fire operations follow a pattern, so the same things happen at each fire. There is always an accurate on scene report, an established water supply, an attack line and a backup line. Exposures are protected, the basement checked, searches performed and reported, and the utilities shut off.

Second Engine

At this point, the second engine company arrives at the hydrant and supplies Engine 1. Next they check the rear of the building, reporting:

Engine 2 on the scene side C, smoke showing from the rear on the first floor.

After checking the basement they report:

Engine 2 to Command, the basement is clear.

This message assures the first floor crews that they are not working above hidden fire. Next the second engine notifies Command:

Engine 2 backing up the first engine's attack line on the first floor.

The backup line is very important because it provides protection for the initial attack crew, increases knockdown power and ensures against burst lines or other problems that may arise. The safety of the attack crew won't depend on a single hose line.

Third Engine

Engine 3, the third arriving engine company lays a large diameter supply line to the front of the building. They do this because they are listening to the fireground radio channel while responding and realize that a lot of water may be needed. Since the backup line is already in place, they notify Command:

Engine 3 assuming the Exposures Sector in the fire building on the floors above the fire.

The IC acknowledges their message and assigns an alternate radio channel to the Exposures Sector. It is best to add another fireground channel as new sectors are established, so that existing channels will be unaffected. The new channel helps reduce the radio traffic.

As Engine 3 enters the building, the residents tell them that people are trapped above the fire. Command is notified.

Command

The philosophy of the IC is to have enough companies to accomplish the tasks and an adequate reserve for unforeseen events.

Command receives the message of people trapped. Considering the life hazards, the current company assignments, and the potential for fire extension, the IC calls for a second alarm, as well as a medical task force comprised of three medical units that respond as a group. These companies are told to report to a staging area which is established several blocks away. The IC then contacts Truck 2 (second due) and assigns them to search and rescue above the fire, reporting to the Search and Vent Sector (Truck 1.) Both Truck 1 and Truck 2 acknowledge that they hear the assignment.

The officer of the first company to arrive in staging becomes the Staging Officer. All responding companies stay off the air. Upon arrival in staging, the driver of the air supply truck automatically becomes the Logistics Officer, and the paramedic becomes the Medical Sector Officer. Both are given vests and checklists outlining their job duties.

Command designates two firefighters from the staging area as aides to Command. They assist with the command chart and radio, and relay information from the pre-plan to the IC. This building has been pre-planned as a target hazard, having the potential for major life and property losses. The pre-plan includes the building's shape, dimensions and exposures. The building is larger than it appears from the outside and has been remodeled several times— both of which spell trouble. The first floor is occupied as a hardware store that contains some hazardous materials. Since the store contains hazardous materials, a hazardous materials company is called to respond to the staging area. The pre-plan sketch shows that the water supply is limited on Main Street, but that Third Street has a large main.

Since the IC has aides to help, he has time to think beyond what the incident initially appears to be to what could happen, playing the "what if" game. Since there is a working fire on the first floor, the IC considers "what if" the fire extends to the upper floors, or into the exposure. If that happens, the IC will need much more help and an adequate reserve in staging. The philosophy of the IC is to have enough companies to accomplish the tasks and an adequate reserve for unforeseen events.

The IC recognizes that there are several dangerous situations at this fire. First, the Attack Sector is operating in the maze of a dark, smoky and cramped hardware store. Second, search crews are operating above the fire. The truck company dispatched on the working fire becomes the rapid intervention team (RIT), relieving the two members assigned when the attack began. Whenever crews are involved in mask operations, the RIT is standing by to rescue firefighters, if necessary. The additional chief who was dispatched on the working fire arrives and becomes the Safety Officer.

THE RIT TEAM STANDING BY
Courtesy of Fairfax County Fire and Rescue,
Fairfax, Virginia

Positioning The Help Companies

The second alarm brings three more engines, a truck, another chief and a canteen truck. The Staging Area Officer issues ICS vests to the Sector Officers as they are assigned. Command alerts Staging that when second alarm engines are assigned from staging, they should lay large diameter supply lines from hydrants on Third Street since the pre-plan shows that this street has a large main. Second alarm engines are also told to supply the first alarm engines that are close to the fire to maximize the water supply.

Command requests an engine and truck company from staging to cover the building on the right side (Exposure D) and assigns them to the same radio frequency assigned to the Exposure Sector. Command confirms this assignment with the Exposure Sector Officer. The chief in Staging is assigned to command the Attack Sector, because control of the fire is a critical consideration.

The Command Chart

The following sectors have been established: Attack, Search and Ventilation, Medical, Exposures, and Staging. The command chart is used to keep track of all the companies on the scene, both for firefighter safety and proper incident management. The IC uses the task check-off system to track what has been accomplished and what remains to be done, to ensure that all the bases are covered.

COMMAND CHART

ADDRESS   224 MAIN ST

SIZE/CONSTRUCTION   30 X 40 ORDINARY

OCCUPANCY   HARDWARE STORE AND APARTMENTS

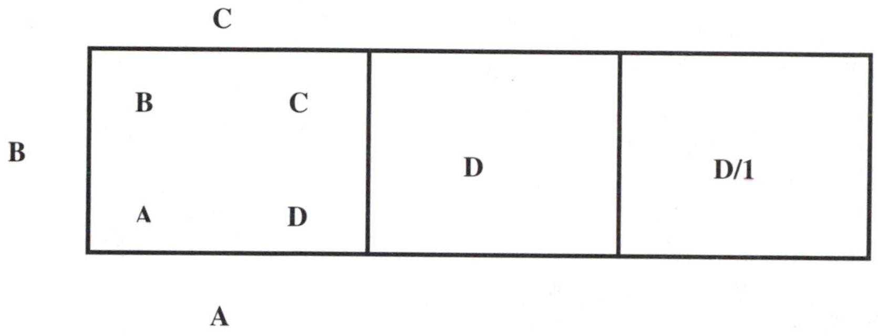

C

B C A D	D	D/1

B

A

STAGING

E 5

E6

E7

M2, 3, 4

ATTACK	VENT/RESCUE	EXPOSURES	MEDICAL	LOGISTICS	SAFETY
C2	T1	E3	M1	AIR 1	C1
E1	T2	E4	A2		T4 – RIT
F2		T3			

BACKUP LINE	(X)	SAFETY ( )	AIDES (X)
EXPOSURES	(X)↑ (X)↔	GAS	( )
BASEMENT CHECK	(X)	ELECTRIC	( )
ROOF	(X)	AIR TEST	( )
PRIMARY SEARCH	( )	SALVAGE	( )
SECONDARY	( )	ATTIC CHECK	( )

Calling For Help

The staging area is rapidly being depleted as the incident escalates. Since a full staging area can be the IC's ace in the hole, the IC calls for a third alarm to respond to the staging area. The IC is thinking big, knowing that his fire department is a part of a regional Dispatch Center that has many resources. This alarm brings another three engines, a truck company and another chief.

Changing Command

The incident operates with a command team who work together to manage the incident.

At this point, a higher ranking chief arrives, he is briefed and command is exchanged. The former IC now becomes the Operations Officer and continues much as before. The Sector Officers are notified of the change and that they now report to the Operations Officer. The incident now has a command team, consisting of the Operations Officer and the new IC working together to manage the incident.

The IC now has time to sit back and look at the big picture, the overall strategy and tactics. If another senior chief arrives and assumes command, both chiefs remain at the command post after the new IC arrives. You now have three chief officers at the command post working together to manage the incident. Access to the Command Post is limited to avoid overcrowding.

TEAM CHIEFING
Courtesy of Alan Etter DC Fire/EMS

Communications

Information is exchanged while radio traffic is minimized.

The IC talks mostly to the Operations Officer. The Operations Officer talks to the Sector Officers. Within Sectors, Sector Officers talk mostly to company officers, preferably using face-to-face communications. This system reduces the volume of radio traffic and maintains the integrity of the command system.

A command channel is established for the chief officers to reduce radio traffic on the fireground channel. The Exposure Sector is already on a separate frequency, so this incident is being managed on three fireground frequencies. The IC uses an aide as the Communications Officer to monitor all the channels for coordination and to make sure that no messages are missed. Cellular phones are used to communicate between Staging and Command to further reduce radio traffic.

FIREGROUND FREQUENCY CHANNEL 2	EXPOSURES FREQUENCY CHANNEL 3	COMMAND FREQUENCY CHANNEL 4

RADIO CHANNEL ASSIGNMENTS

At this point, Dispatch notifies Command that they are now ten minutes into the incident. Command acknowledges, gives a brief progress report and conducts a quick accountability check through the Sector Officers. All personnel are accounted for. They begin accountability checks early in the incident because that's when they're needed most to prevent firefighter deaths and injuries. Additional checks are made at ten minute intervals thereafter.

Sector Reports

Sectors report every five minutes; reports include personnel accountability.

The Sector Officers give progress reports to Operations every five minutes, without Operations having to ask for them.

The Attack Sector reports: *Engine 1 and Engine 3 are attacking the fire, progress is slow and difficult. We are PAR.* (The Personnel Accountability Report shows all members in the sector are accounted for.) Sectors routinely give these accountability reports along with progress reports. Reacting to this message, Operations assigns another engine company from Staging to work in the Attack Sector.

The Search and Vent Sector report that they have found several victims on the second floor, they will be bringing victims out the front and they are PAR. Although the building is laddered, they will use the stairs because inside rescues are safer than ladder rescues. Command coordinates the rescue with the Medical Sector to see that they will be prepared for the victims.

The Attack Sector reports that a member of Engine 1 is running out of air, so the entire engine crew is leaving the building. Each company enters together, works together and leaves together. Reacting to this message, Operations assigns another engine company from Staging to replace Engine 1; and Command establishes a Rehab Sector.

Exposures are giving a progress report to Operations when suddenly the radio barks ***Emergency traffic!*** The Exposures Sector falls silent. *Emergency traffic, Search to Command,* comes the muffled message through a mask, *we need help getting the victims out of here.* The IC tells the

11

rapid intervention team to help the search team and they instantly go into action to help the search crew with the rescue. This shows why it is important to have the RIT available and the value of using emergency traffic messages to clear the radio channel for critical messages. Command then replaces the RIT with another company.

The Attack Sector reports: *We have fire in the partitions in the fire building, and we need another engine and truck company.* Command reacts by assigning the requested companies from Staging and requests Dispatch to send an aerial tower. The IC recognizes that in any doubtful operation, the presence of a well-positioned tower can be a good alternate plan.

THE AERIAL TOWER CAN BE AN ALTERNATE PLAN

The IC prepares for a possible master stream operation which may require positioning the tower and ladder pipes. An aide is sent to scout truck company positions and a water supply for the truck companies.

Because of the deteriorating situation in the fire building, the Exposures Sector is becoming more critical. Command assigns a chief officer to assume command of the Exposures Sector, relieving the company officer from Engine 2. Exposures are notified that the pre-plan indicates several renovations have been done to the exposure building in the past—not good news.

More messages from Sectors to Operations:

Search to Operations, primary search complete, three people removed from the building.

Exposures to Operations, exposure D has heavy smoke, but no fire.

Vent to Operations, electricity has been shut off in the fire building. PAR

Search to Operations, secondary search complete. Results negative. PAR

Dispatch notifies Command that they are now twenty minutes into the incident. Command acknowledges and calls the Exposures Sector for an accountability check. There is no need to call the other sectors because they all have reported PAR within the last few minutes as part of their normal radio transmissions.

The Attack Sector reports: *The fire is in the attic of the fire building.* Command makes sure that the Exposures Sector copied the message. Fortunately the IC has already started preparations for a master stream operation.

Changing From Offense To Defense

When in doubt pull them out.

The IC is now faced with a deteriorating fire situation, fire on all floors. A secondary search confirms that all the occupants are out of the building. Time is passing, and little progress is being made. The risks are increasing, and now there is some doubt about the stability of the building.

The IC decides to pull the plug, "When in doubt pull them out." An orderly, safe withdrawal will make the accountability check easier. Safety is advised of the decision and told to establish a collapse zone with cones and barrier tape. The word goes out: *Operations to all sectors—abandon the building, we are going into defensive operations.* After the radio announcements have been made and acknowledged, apparatus air horns are sounded for ten to fifteen seconds.

Withdrawing companies leave together and report to their apparatus where the company officers conduct a head count of each crew. Sector Officers account for their companies to Command. This accountability check is required when changing to a defensive position, to verify that everyone is out of the fire building.

The fire is still out of control, and Exposure D (the right side) is attached to the fire building, so additional companies are assigned to protect this exposure.

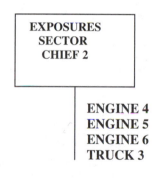

THE EXPOSURES SECTOR

The IC orders that the fireground be divided into sectors A, B, and C to cover all sides of the building, and assigns a chief to each sector. The Sector Officers are given the job of assembling their own sectors by determining which companies are already in their sectors, their sector needs, and they report any overages or shortages. Command then makes the necessary adjustments, making sure that each sector is up to strength. Command tells the Sector Officers that the tower and truck companies are being positioned and supplied with water and that everyone is out of the building. They are to begin defensive operations when ready.

Within a few minutes (because the IC anticipated a defensive operation), heavy streams are working on the fire building. Exposures report that all is well, and the Safety Officer reports that the collapse zone is in place and being monitored. The Rehab Sector is rotating firefighters into a rest area, where their vital signs are evaluated. The canteen provides food and drink.

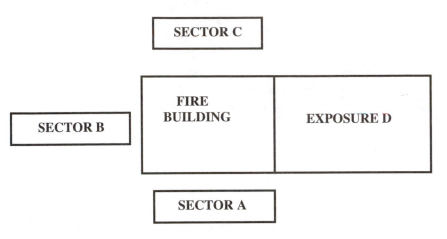

THE FIREGROUND DEFENSIVE SECTORS

De-Escalation of The Incident

After the fire is controlled, and the exposure saved, Command sends the Safety Officer and a select crew to conduct a safety check of the fire building. They look at the stability and safety of the fire building. Unsafe areas are made safe or taped off, and no companies are allowed inside until it is safe.

Exposures reports that: *The smoke in Exposure D is clearing up.* The Safety Officer quickly takes air monitor readings. When the monitors show safe concentrations, the announcement is made: *Operations to Exposures: air monitoring negative, all clear, okay to take masks off.*

The IC does not return fire companies too quickly because unexpected things can happen. Overhaul is often a bigger job than it appears, especially when the troops are worn out. The philosophy is to take it slow and easy and make sure that nobody gets hurt. When the IC begins returning companies, he starts by releasing the exhausted, first arriving companies.

Post Incident Analysis (PIA)

Every incident is an opportunity to learn and improve operations.

A PIA is held the next day, because if too much time goes by, memories fade. The PIA shows what went right as well as what went wrong. Some mistakes were made, but over all, this fire went well. The building was vented, the occupants rescued, the exposure saved, and, no firefighters were hurt. The results of the PIA are written and remain a record for future use.

Don't you wish that your firegrounds operated like this? The information in this guide can get you there.

2

PHILOSOPHY OF EMERGENCY OPERATIONS

"Nothing can stop the man with the right mental attitude from achieving his goal." - *Thomas Jefferson*

Our philosophy is very important because it is our guiding light. It reflects our goals and how we do business at the emergency scene. We need to think correctly to be able to resolve the emergency situations that we face.

Years ago the fire service was brainwashed into the philosophy that our job was to save life or property at any cost. I remember talking to a firefighter many years ago about the tragic loss of four firefighters who died in a building collapse. I wondered why these firefighters died in a vain effort to save an abandoned building. His answer was chilling. He said, "That is what we are paid to do." That's the way it was in many fire departments in years past. Taking reasonable risks to save lives is valid, but dying for property has never made sense.

Don't die for property.

Darnell And Dicky

Steady, experienced, and calm is much better than unbridled enthusiasm.

When I went to Heavy Rescue 2 as a newly assigned captain, two of the firefighters assigned to my platoon were Darnell and Dicky. They were opposites. Dicky was small, and young. Darnell was burly and had been around forever. Their philosophies were different. Dickey prayed for fires. Darnell prayed for no fires. When we got to the scene of a fire Dicky was just raring to go, while Darnell would wish that he was on day off. These two opposites would then go inside the burning building as a search team. I was always very glad that Darnell was there to look after Dicky. Darnell never got excited, and that is a big plus in fire fighting. You could count on Darnell to do the right thing and get back out alive with Dicky.

Don't Lose Your Cool

Don't become part of the problem, become part of the solution.

It was a working fire. We had it knocked down, but it became critical to start pulling the ceilings. I started yelling for hooks and personnel. I was a new lieutenant. I was young, dumb and excited. I was (this is embarrassing to admit) standing in the fire area screaming repeatedly that I needed hooks. Interestingly, no one paid much attention. I guess it's considered somewhat normal to lose your mind at fires. I caught hold of myself, soon the ceiling was pulled, and everything was all right. Later, in thinking about my performance, I vowed that it would never happen again. I realized that, though the work had to be done quickly, screaming didn't help. Self discipline is critical to firefighter safety.

Fire is unforgiving and losing your head can be dangerous. You soon realize that by becoming excessively excited you become part of the problem, not part of the solution. You have to be able to count on yourself, because others are depending on you. Being cool when fires are difficult and getting worse requires practice, discipline and experience. Sometimes you have to talk to yourself, saying things such as, "I didn't cause this, and all I can do is the best I can under the circumstances." Take time to be safe, do your reasonable best and don't take stupid chances.

FIREFIGHTER PERSONALITIES

Firefighters are a diverse, interesting and sometimes lovable group. The following observations are offered to help you better understand and work with your fellow firefighters.

The Leaders

These are the folks you wish that you had more of. They are the ones who recognize problems and try to improve things. Leaders are mentors who impart their wisdom to less experienced members. Good leaders lead by example and deeds, not just by giving orders. Do not confuse leadership with rank—they are not necessarily the same.

I was once part of an investigation of a fire where operational problems led to firefighter deaths. During the interviews we found that very few firefighters or officers had activated their Personal Alert Safety System (PASS) devices before entering the burning structure. The exception was one company where all four members had activated their PASS devices. Why did they have their PASS devices activated when almost nobody else did? Their answer was that Captain Rodgers insists that they do it. Captain Rodgers wasn't with them that day. They did it anyway because their captain was interested in their safety and ran a tight ship. That's outstanding leadership: when your troops do the right thing even when you're not there because of your example and influence.

Dinosaurs

Dinosaurs live in the past and are against change. The old way is good enough for them, and they will argue "that won't work here; we're different." The dinosaurs are not necessarily senior people; the problem is more attitude than seniority.

The Kamikazes

These folks are scary. They think that they are invincible. When they see fire, they go full speed ahead. Fire in an abandoned building, Charge! When in doubt, Charge! They push and shove and shout. They scoff at SOPs and think that taking foolish chances is macho. These firefighters will charge blindly ahead while air horns and radios are screaming retreat. Sometimes the kamikazes get promoted to become Lieutenant Kamikaze, which is really scary. It is bad enough for a kamikaze to take unwarranted risks, but it is far worse to risk the safety of the firefighters he is responsible for.

Followers

Most members of the fire service are in this category. These are your good steady people who are influenced by both leaders and dinosaurs.

The Tell It Like It Is Folks

We call him Captain Straight Talk. He is a solid well-rounded firefighter. The kind of guy that you'd want to come get you if you were trapped. Whenever I want the straight scoop on what really happened or an honest opinion on something, I ask Captain Straight Talk. He is totally honest and brutally frank. If he is not happy with something, he minces no words. He tells you straight out, chief or not. I love him for it. There aren't many people who will be totally candid with the boss. Most officers will tell their chief what the chief wants to hear or will be evasive and dance around the question. Some captains are known as Captain Suck Up. These folks want to be a chief and will say or do whatever it takes to become one. Not my Captain Straight Talk. He is about as subtle as a bulldozer. The Captain Straight Talks are gold because they can help you to improve your operations.

Dirty Little Secrets

The fire was coming out an upper floor window. A firefighter pulled the longest pre-connect, but the captain said no, take the shortest line. The firefighter dropped the long line and started in the building where he soon ran out of hose. It took time to extend the line. When the water finally came, the line burst. By that time, another company had put out the fire. Later, the captain, who was not on the line where he should have been, said that he couldn't understand why the line burst because he had opened the gate on the pumper very slowly. Later the captain was promoted to chief.

All fire departments (both career and volunteer) have members who don't belong in the fire department, but few try to do anything about it. It's a problem that we all know exists, but I have never seen it in print. It is often viewed as an in-house joke that's snickered about in private. In larger departments, sometimes the misfits tend to wind up at some outlying station that either attracts the less competent, or is used as a dumping ground for misfits, troublemakers, and malcontents. Too often, the problem people are simply moved around so that somebody else will have to deal with them. There is a serious reality gap here. We claim to be into modern management with all the appropriate buzz words: management by objectives, total quality management, and assessment centers. But we continue to have the same personnel problems.

Solving this problem is not easy, but it can be done by implementing a more stringent entry screening process, a closely monitored probationary period, and taking politics and nepotism out of the process as much as possible. When we have problem firefighters, we should make every effort to bring them up to snuff.

Beware Of Captain Clean

The newly promoted captain came to us all spit and polish. Dust was his enemy. He liked to clean the undersides of chairs. He wanted to alphabetize the spices in the kitchen cabinets. He checked the bathrooms to make sure that all the toilet paper rolls had the loose end on the outside. His "efficiency" was a constant pain in the rear.

I was waiting to see how efficient he was when we caught a working fire. After several weeks of killing dust balls by the thousands, we responded to a working fire on the third floor of a large commercial building. We hustled our pre-connect up to the third floor and were met by a wall of smoke. Visibility was zero. His efficiency stopped there. We crawled in, found the fire, put it out and threw it out the window, all unsupervised. We never saw or heard a word from him. As they say in Texas, he was all hat and no cattle. He stayed with us for a while after that, but his credibility was shot. It wasn't too long before he transferred to a slower station where there were more dust balls than fires.

Some of you probably know some of these officers who have a fascination for cleanliness and small details because that's all they're good at. They are efficient as hell until the bell sounds. Then they cease to be leaders, and become cheerleaders — Go get it guys! Cleanliness nuts are rarely very good on the fireground. I don't mean that good officers are sloppy, just that most good officers do everything well, including keeping things clean.

New Officers

To be a good officer you must be competent, fair, and always try to do the right thing.

It is usually a difficult transition going from firefighter to officer. Often even your name changes. You are no longer Harry. Your name suddenly becomes Lieutenant. You used to be responsible only for yourself, but now you are expected to lead and show the way with the lives of others depending upon your actions.

Most new officers are uncomfortable in their roles. They are under the gun, and everybody knows it. People are looking to see how well they handle themselves. The people doing the looking can be their friends or peers, so they are under considerable pressure. They have to adjust to new names, new titles, and new jobs. They have to be in the spotlight while wanting everybody to love them. Almost all of us want to be liked, and loved is better.

It is easier to get along than it is to lead. Too many officers just slide along trying to be popular. A wise old chief once told a mint green new officer, "When some people call you an S.O.B., it's a compliment." When you are doing the right thing, most of your people will appreciate it. Some may not like it because it upsets their routine. Years later the same officer said, "I found this to be one of the best pieces of advice that an officer can use when facing a tough decision. It takes a lot of the fear of offending people out of the equation."

Many promotional exams in the fire service use assessment centers designed to measure a candidate's ability to resolve serious and complex personnel problems. One thing the tests don't take into account is that good fire officers prevent personnel problems from happening in the first place. They monitor day to day activities, recognize developing problems, and take action to prevent the problems from becoming serious. Take care of the little problems, and the big ones will take care of themselves.

Another hallmark of a good officer is consistency. Most folks would rather work for a boss who acts the same day in and day out. They know where that officer is coming from and what to expect. They don't have to guess what mood the boss is going to be in today. That beats the "I'm your pal today and Hitler tomorrow" type of boss.

To be a good officer you must be competent, fair, and always try to do the right thing. You must look out for the troops. They need to feel that, if push comes to shove, you will be there for them. It comes down to trust. The troops need to be able to trust you, and that trust is earned over time. As an unknown author said, *"A person is not given integrity. It results from the relentless pursuit of honesty at all times."*

Going from a company officer to chief can be difficult. Most departments have had the experience of an excellent company officer making a mediocre chief. On the other hand, some mediocre company officers have made excellent chief officers. The skills needed to do the jobs are different. The company officer is usually a hands-on position, in the middle of everything. The command role demands the opposite. Those in command must delegate, making sure that things get done, not doing them or becoming involved in details. Some chiefs never really make this transition.

CHANGE IN THE FIRE SERVICE

If you want to improve operations in your department, you have to be ready to encounter resistance to change—in individuals and in the department as a whole. Overcoming resistance is difficult, but worth the effort.

Making Changes

People often don't like changes, so if you want to change things, you need to sell the idea.

It is difficult to change the way we operate on the fireground. The big wheels tend to think that the troops will comply with any directives that come from on high. My experience is that it seldom works that way. People often don't like change, so if you want to change things you need first to sell the idea. Everyone won't buy into it, but if the idea is sound, the change will fly. It is very important for the troops to be involved in the development process for the new procedures or changes. When we developed new SOPs in the Washington, D.C. Fire Department, we involved everyone in the process. **If a firefighter made a good suggestion we adopted it.** This decision was probably the key to our success. Soon the troops viewed it as "our project," not something that the chief was trying to ram down their throats. To make a long story short, the new SOPs were very successful because of the support of the rank and file.

INVOLVE THE TROOPS IN THE CHANGE PROCESS

Attitudes

You can complain about change or adjust to it. Adjusting is more effective.

There is often a lot of frustration among the senior officers and senior firefighters. They complain a lot that they want to "Go back to the good old days" and "It just isn't fun anymore." Funny, but during my early years in the fire service, I remember the old-timers complaining about how the fire service had gone down the tubes. Now those times are looked back on as the good old days. Everything changes in this world, and we need to adjust to it. Some things are better, and some things are worse than they used to be. One thing that hasn't changed much is that most of our recruits today still have stars in their eyes, just like the rookies did in the good old days.

DEPARTMENTS HAVE THEIR OWN PERSONALITIES

Fire departments have personalities just like people. Some are boastful, "We are big, tough, and the best." Pride is great up to a point but, when a departments members have a shared belief that they have the greatest department, it is very hard to make changes or learn from other fire departments. Some departments are the opposite: low-key, hiding their light under a bushel. They often have some really good procedures, but they normally won't brag or even tell you about their practices unless you ask, and they are often looking for better ways to operate.

My Department Is The Best

It is very difficult to improve when the only operation that you know is your own.

The vast majority of firefighters have worked only in their own departments, and have little idea of how other departments operate. It is very difficult to improve when the only operation that you know is your own. How can you conclude that your operations are superior when you don't know what others are doing? You can't, but it is easy to be smug and comfortable in your ignorance. The truth is that all fire departments do some things very well, and all have room for improvement. The fire service is coming closer together, but it is a very slow process. Too often, departments reinvent the wheel when other departments already have the information that they need. The challenge is to expand our horizons so that we share ideas and learn from each other. Information can be exchanged through trade journals, the internet, and classes and professional organizations.

Some progressive departments have set up programs to exchange personnel with other departments on a regular basis. This sharing of information and attitudes helps both the firefighters and fire departments to expand their horizons.

MORALE

Morale is great in many fire departments because we have a lot of really good people who do important work in teams. Working together under difficult and dangerous conditions builds morale. Bonds also form through eating and socializing together. Morale improves when we care about our fellow firefighters and when the troops believe that the brass cares about them.

I once visited FDNY's Rescue 1 in Manhattan. They had recently come from the funeral of a retired member of their company. They told me that the firefighter had been retired thirty-two years, and that many had gone to the funeral. They said that he was one of them and that they wanted to be there for his family. The family was overcome by this turnout. Only in the fire service do you find this sort of camaraderie.

On the other hand, some companies or fire departments don't seem to get along very well. The reasons for poor morale are the opposite. When the brass doesn't seem to care, when work is strictly business, or people are treated unfairly, things go down hill fast. Losing out on deserved promotions, either by vote in the volunteer departments or by politics in career departments, has a devastating effect on morale.

**FIREHOUSE MEALS
IMPROVE MORALE**

SUMMARY

People often don't like changes, so if you want to change things you should sell the idea. If the troops view it as "our project," the changes will probably be successful. Most members of the fire service are sincere, steady people.

By creating a good entry screening process, using a strong probationary period and taking politics out of the process, we can maintain high competency standards.

The challenge for fire departments is to expand our horizons so that we share ideas and learn from each other. Setting up exchange programs with other departments will help the learning process.

3

PREPARATION

"We are what we repeatedly do. Excellence, then, is not an act, but a habit." - Aristotle

Preparation is the key to the success of our emergency operations. This chapter provides guidelines to help you improve the preparation levels in your fire department.

Check And Double Check Equipment

Check your equipment faithfully and regularly as if your life depends on it, because it does. Develop a ritual to check everything thoroughly and consistently. When you are responding from home, this may be difficult. If you are unable to check your equipment before responding, check it on the fireground before using it. Of course, this is easier said than done. There is a lot of pressure to hurry, to be the first one "on the knob," to put the fire out and save the day. Nobody wants to be known as the one who is always outside fiddling with the mask. The surface answer to this problem is training, but it goes deeper than that, into people's attitudes. Macho attitudes should never get in the way of safety, but too often it does. Checking equipment is not glamorous, but very necessary. Finding out that your mask doesn't work properly when you are in a dark and smoke-filled building increases the already high danger factor. Another common problem is finding that your gloves are missing just as fire attack begins. Do you slow down the operation, or take a chance working without gloves? (a really bad idea.) Finding out that your water supply tank is empty when your attack crew calls for water is another example of failure to check your equipment that can lead to firefighter injuries. There is no excuse for not checking your equipment.

Never Assume A Call Is Routine

All calls have the potential for injury or death for the responding firefighters.

Most incidents that fire departments respond to are minor. Firefighters learn to expect certain types of problems in certain situations. As a result we can become lax, and develop a "smells and bells" mentality. It is easy to fall into the rut of routine. We begin to think "it's going to be another false alarm" or, "the report of smoke is probably just an odor." We must keep alert because all calls have the potential for injury or death for responding firefighters. Firefighters can get to where they feel invincible, but they don't realize that there is a fine line between what they have seen and what can happen. In the words of one firefighter after a fellow firefighter fell through the floor and died, "The fireground went from almost placid nothing to total terror within seconds."

Some years ago a firefighter died after falling through a hole that had been covered by a piece of sheet metal. The fire involved one room in a store, and the fire had mostly been knocked down. It appeared to be a routine fire. Even after the firefighter fell partially into the hole, it was not apparent that he was in serious trouble. Following the death of Firefighter John Williams of Rescue Squad 1, Captain Thomas Gregory, District of Columbia Fire Department stated, "This was almost a routine fire. Just a room and contents. Almost routine."

The Downtown Snakebite – A Bizarre Report Just Might Be Accurate

There was no way any snake could exist there.

The crew of Engine 16 was joking and laughing after they were dispatched for a report of a snake bite in McPherson Park. The park is very small and surrounded by high-rise buildings. There was no way any snake could exist there, and a poisonous snake was out of the question. It was late at night, and it would be unusual to find a person in that commercial area, even if there was a snake.

The joviality ended when they arrived. There was indeed a snake bite victim who was in desperate need of assistance. The victim had broken into the reptile house at the zoo and stolen an exotic poisonous snake. He had put the snake in a black plastic garbage bag and boarded a bus to take it home. When he'd climbed off the bus, he'd carried the bag with the snake over his shoulder. The snake had struck through the bag and had very nearly killed him. The victim recovered after receiving an antidote flown from Africa.

The Elevator Nightmare – Ask Questions First

One afternoon we responded to help some people trapped in an elevator. Upon our arrival the power was obviously off, and building maintenance people were working with flashlights trying to figure out how to extricate the people from the elevator. The elevator was located in a blind shaft which meant that there were no elevator doors near them. We placed a ladder down the shaft to the top of the elevator. Then we used a step ladder to assist the people to the car roof where we guided them up the ladder and out of the shaft. It was a lengthy operation since there were four or five people in the elevator car.

As soon as we were finished and started to pick up our equipment, the power suddenly came on and away went the elevator. Startled, we asked the maintenance people who had turned on the power. They replied, "The power was never off, there was an electrical outage in the area." I could close my eyes and see the elevator moving down the shaft with firefighters and civilians on ladders on top of it. It was a close call and another learning experience.

Assumptions and laziness can get you killed. During each call continue to use every piece of information you have and consider possible causes, interventions, and effects. Making snap judgments may cause you to overlook critical safety information.

THE ELEVATOR NIGHTMARE

 There is no such thing as a routine call.

24

Prepare To Find Fire

When the ho hum investigation discovers a real fire, panic often follows.

Companies responded to a report of a fire in a basement clothes dryer. Ho hum, only another routine call. Nothing was showing upon their arrival, that was normal. The first engine went in unprepared, as usual for them. There was smoke on the first floor, not a good sign. They made their way to the basement where the dryer was located, and it was mask-up time. The engine crew crawled around the basement without a hose line looking for the fire. Meanwhile, there was an engine full of hoses and water sitting out front.

The engine crew found clothes blazing on top of the dryer. They opened the dryer and a fireball erupted from the opening. Now they really needed water in a hurry. Not having any water, they started looking for a source. They searched around the laundry room and found an old pitcher near the sink. Now the firefighters were reduced to running back and forth from the sink to attempt to extinguish the fire. Sound familiar? Unfortunately, many firefighters have had to play catch up because they were unprepared. They put out the fire. Some of the firefighters thought they had done a good job because they had been creative and improvised. Others were disgusted by the mickey mouse operation.

In the fire fighting business, we pride ourselves on getting out of the station quickly, and getting to the scene as rapidly as we can. However when we get to the scene, unless a fire is obvious, we generally investigate. In the army you often have to hurry up, and then you wait. Often, in the fire service, we play "hurry up, and then investigate."

This often means giving little or no thought to a water supply, an attack line, self-contained breathing apparatus (SCBA), or full protective equipment. Usually a few firefighters will go in, prepared to find nothing, and the rest will wait outside expecting to go home. Looking for a fire when you are not prepared is like leaving your bear gun in the truck and going out to look for the bear. If you find a bear, you have to go back to the truck for the gun.

**WHEN LOOKING FOR BEAR,
TAKE YOUR GUN**

Unprepared firefighters are only ready to find little or nothing and return to the station. When the ho hum investigation reveals a fire, panic often follows with a lot of yelling, screaming, and trying to play catch up. Remember, the fire was *supposed* to be there, that's why we were called.

In some fire departments there are two modes: This is another nothing incident, and, Oh my God it's a fire. The "Oh God" operation isn't pretty or predictable. Things often go downhill fast, and important things are often overlooked. Operations that start badly seldom end up well. The philosophy of every member responding should be "This is serious business, we are going to a **FIRE**." Unfortunately, lazy firefighters often get away with being lazy because most fires aren't serious.

The Reality Gap — A Small Fire Screw-Up

It was just a routine one-room fire but, as it turned out there were problems. The first engine charged the wrong line. Then they had a hard time getting water in the right line. There was lots of screaming and yelling. The truck's search was very late because the truck crew was on a hose line. The IC never realized the search was late or water was a problem, and never called for any help. Eventually they muddled through and the fire was put out.

This fire occurred in a career department with reasonably experienced companies. If you had questioned this crew before the fire, they probably would have done everything the right way on the blackboard. The truck officer didn't get promoted through an assessment center by saying he ignores life safety if he can get on a hose line. The chief would have said that he would have called for help because of limited personnel, a large house, and the problems in getting water. The reality was something else. There is often a big gap between what we say we are going to do and what we actually do. Rick Dempsey, former catcher for the Baltimore Orioles said: "You have to play this game right. You have to think right. You've got to do the things that you've talked about and agreed upon beforehand." Rick was talking about baseball, but his approach also applies equally well to the fire fighting business.

A Big Fire Screw-Up

We have all been to fires where nothing seems to go right.

The first engine went on the scene of a fire in a garden apartment screaming that there was fire showing and that they needed an immediate backup line. They were dispatched to a fire, so they shouldn't have been surprised, and a backup line should be SOP. All efforts went into extinguishing the fire located on the second floor of a three-story garden apartment. There was very little ventilation and no exposure protection.

Calling for help was slow and piecemeal. The original dispatch got three engines, a truck company and a chief. It was obvious that they were going to need a lot of help quickly (what they got was a little help slowly.) Ten minutes into the incident the IC requested two more engines. Later he requested another engine then another truck, followed by another request for an engine and truck, followed by another engine. Help was requested in bits and pieces over the next hour.

No staging was used, and no companies were assigned to sectors. As a result, the IC was overwhelmed with radio traffic. When the visible flame was knocked down, they all thought that the fire was over, and there was a lull in communications. Then all hell broke loose. There were cries of fire in the attic. Then they were screaming for hose lines to the roof. (You don't want to know what they did with the lines.) Finally a company was assigned to protect the exposure in the attached building, which had not been checked. Soon they were screaming that they found heavy smoke in the exposure, and they needed lines and hooks. There were never any organized searches; nor was there a critique or post incident analysis. The IC said that he thought that the fire went pretty well. Unfortunately, in some departments, these fires *are* the norm, as if fire calls are meant to be chaotic because that's the nature of the business.

We have all been to big and small fires where nothing seems to go right, so we can relate to some of the mistakes made at these fires. They are all common errors that haunt us regularly. In this age of ICS, command and communications (the radios used in both fires were state-of the-art-800-frequency radios), you would think that fires would not go this badly, but they do. The main reason that bungled fires continue to happen is that we often skate along doing the bare minimum on our routine calls, and when we have serious fires, our bad habits come back to haunt us.

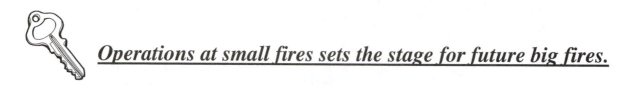

Operations at small fires sets the stage for future big fires.

Always Operate The Right Way – Even If It Doesn't Seem Necessary

Some years ago I was a newly assigned lieutenant with Engine 18, a busy company in the Capitol Hill area. We often freelanced, like many fire departments during that era. This meant that your engine company on the fireground could often do pretty much what you wanted, unless a chief told you otherwise. It wasn't a very efficient system, but it was a hell of a lot of fun, and it fostered competitiveness between the engine companies. It was an "you snooze, you lose" system that rewarded speed and aggressiveness.

The Attack Plan

Engine 18 was my first engine command, and I wanted to do my best and avoid some of the mistakes that I had seen other officers make. My game plan was simple— whenever we responded to a reported structure fire, we operated as if we were the only engine company responding, and assumed it was a working fire. This meant that we laid supply lines and attack lines to where the fire was reported and used full protective equipment, including SCBA. We were going to arrive at every incident fully prepared to do battle with the red devil.

The hard part was to explain this plan to my incredulous troops. Engine 18's pumper operator stopped talking to me. After a few weeks, just about the time my relations with the troops were at the breaking point, we caught a fire that changed everything.

The Fire

The fire was reported in the basement of a high rise apartment, and we were the third due engine company. Upon our arrival, there was nothing showing. The first engine and truck were investigating, and the other companies were hanging around outside. They watched with great amusement as 18 Engine and their crazy lieutenant advanced an attack line into the building and down the stairs. When we got to the first landing, we found smoke, and the first due companies passed us as they bailed out of the basement in a near panic on their way out to get ready for the fire. We simply continued down the stairs and calmly extinguished the fire, which is why we were sent there.

WE WENT IN AND PUT OUT THE FIRE

Preparation Pays

That was a turning point with my relations with the crew of Engine 18, and we continued our aggressive tactics. Over the next few years, we caught a lot of fires, many of which were stolen from less prepared companies. We rapidly developed a reputation for being supermen. We started to hear stories that some of the other companies were getting so gun shy of us, that they were worried about our stealing their fires, sometimes when we weren't even there.

The truth is that we were not any better firefighters than the others. The secret was that we were simply **more prepared** then they were. A successful major league baseball pitcher said once that he doesn't "out ability" everybody else; he just out smarts them because he is prepared. To win you first need the will to be prepared to win. Fire operations are similar. A major positive spin-off of our prepared approach was that the basics of engine company operations became second nature to us. This is the key to getting your people to do the right things the right way at the right time. We avoided most of the common errors, such as pulling the wrong attack line, getting lines tangled or snagged, charging the wrong line, having problems while masking up, or not getting water at the right time and pressure. We were prepared and treated every run the same: layout, mask up and stretch in. We also avoided a common fireground disease known as being "rusty." We became a well-oiled team and put out a lot of fires with a minimum of errors.

 __Always do it the right way. It isn't always necessary, but it will ensure that you will be doing it the right way when it counts.__

Train To Survive

Training is an important part of being prepared.

Too often training has a bad reputation in the fire service. Mention training, and some will roll their eyes, as they recall boring uncomfortable sessions where they were read to. We have a new generation of firefighters to train. The MTV generation doesn't want Lawrence Welk instruction. Teaching needs to catch up with the times.

The following observations are offered as ideas that you may be able to use to improve training in your fire department.

For effective fire service training, choose important subjects taught by good instructors and actively involve everyone in the learning process. Training must be practical and relate to the real world. Nobody is going to fall asleep during a finding-lost-firefighter drill. Everybody is going to fall asleep during classes where instructors read from departmental by-laws or from the text "Improved management techniques through programming."

The fire service has become so complex that trying to keep up with all the different aspects can be overwhelming. We must focus on the basics, such as using SCBA, making sure that everybody can get water in their attack lines and learning your area (you can't do anything if you can't get there.) Train on the problem areas in your department, and don't take anything for granted. Following a serious failure to get water at a fire, a large fire department sent several pump operators to a retraining session. Assume nothing.

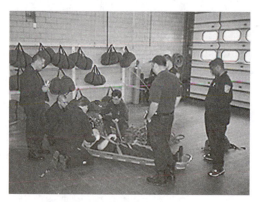

TRAINING IS IMPORTANT

Visuals are important. Watching videos of your own or other fire departments fireground operations can be a learning (and sometimes amusing) experience. Stop the tape periodically to discuss what went right or wrong. Play back important scenes following the discussion to reinforce lessons.

Simulation is an excellent learning tool. It doesn't have to be fancy. All you need are some portable radios, some visuals, command charts and a few role players. Simulation is effective because it is learning by doing, and valuable because few fire departments have enough major incidents to stay sharp with on the job training. I once took a regional fire officer command course with many simulation exercises. About three months later, I was the IC at a major building fire that involved about twenty-five fire companies. The incident went remarkably well, and during the fire I had this odd feeling that I had done this before. Then it

29

dawned on me that I had done it before with my earlier simulation exercise. Simulation really works.

USE GREAT INSTRUCTORS

When possible, choose instructors who love the subjects they are teaching. When instructors are mandated to drill in areas where they're not comfortable, training often suffers. Instructors should be chosen by their expertise, and not their rank. Reach out to people in other departments and from other shifts. Using different instructors keeps training fresh.

Instructors should use less lecture and more class participation. They should ask probing questions reflecting real life situations to stir the students' minds. If the instructor is the only one talking, the training session is dead. If you show someone how to do something, they will pay attention, but real learning takes place when they actually do it. Firefighters like most adults, learn by doing. This dynamic learning avoids the "I have to sit through this to get my ticket punched" mentality.

Industry often sends people on retreats, so they can learn in a completely different environment. Most fire departments lack the resources for this, but variety and interest can be added by getting out of the fire stations. Get out and look over the rail yard. Arrange for mutual-aid disaster drills with neighboring fire departments. Go to the junkyard to cut up cars, or hold high-rise drills when buildings are closed. You get the idea. A good instructor is only limited by his or her imagination. Don't just talk about how to do things—do them.

Every run is a training session. Minor incidents are training grounds for major incidents. Try to do the basics properly on each run. Don't get careless or sloppy. After incidents, talk about what happened and how things can be improved. Incident reviews are also helpful in assisting the new recruit to learn the ropes.

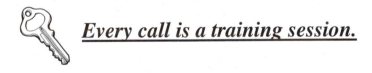

Every call is a training session.

Teach How To Avoid Being Lost Or Trapped

Engine 21 was first on the scene of a fire in a large warehouse. Sprinklers had extinguished most of the fire prior to their arrival. Using a 1¾-inch hose line they quickly extinguished the remaining fire. Smoke conditions were heavy however, and one firefighter and his partner left the security of the hose line to start ventilation. The pair became separated, and soon one firefighter was alone and running dangerously low on air. He did not call for help as he wandered through the cavernous building looking for a way out. Finally, he reported himself trapped with his last breath of cylinder air. A search was immediately started. A rescue crew randomly forced open one of many overhead bay doors and unbelievably found the downed firefighter on the other side. He had stopped breathing. Fortunately the firefighter survived, but it was a very close call. This particular firefighter ran through many red lights on the fireground, including:

- Failure to recognize increased danger when operating in an unusual environment

- Failure to stay with the safety of a hose line

- Failure to have any sure way out

- Failure to stay with his partner

- Failure to call for help as soon as he became lost

We must teach our people how to avoid getting into the soup, and how to get out if they do. Unfortunately, most training focuses on hardware use. If you operate your nozzle but not your head, you can be in trouble real fast. If you follow street smart procedures, you dramatically decrease your chances of becoming lost or trapped.

The following are the surest ways to stay out of trouble on the fireground:

- During size-up, look at the windows for security bars, air conditioners or casement windows that could impede your escape

- Pay close attention to the location of the fire

- Consider the risks against the gains

- If you are not in a residence, be extra careful

- Go in as a team, stay together and exit together

- Closely monitor the amount of air remaining in your cylinder

- Stay oriented by using a hose line, wall or a search rope

- Avoid situations where you are not protected by a charged hose line

- Avoid stumbling around in heat and smoke by venting early and often

- Break out the windows from the outside

- Break out the windows as you advance inside

- Put out the fire

- Follow your SOPs

- Always know where your escape routes are

- If you are not making progress and increased ventilation and more water doesn't help, consider retreating to a safe area

- Remember that you are ultimately responsible for your own safety. A back out order doesn't have to come from Command

TRAIN TO SURVIVE

Teach Firefighter Survival Training

The proudest ship in the navy drills regularly on abandoning ship. Every sailor on every submarine knows how to escape from a stricken sub. It's amazing that we send people into burning buildings but often don't teach them how to get out under emergency conditions.

Some progressive fire departments teach firefighters how to survive when lost or trapped in a burning building by training on the following techniques:

- Calling for help as soon as you realize that you are in trouble

- Manual activation of the emergency PASS device

- Closing doors between you and the fire

- How to conserve air when lost

- How to do air sharing/buddy breathing

- How to squeeze through narrow spaces using an SCBA low profile maneuver

- How to perform an emergency bailout using a personal rope

Teach firefighter survival skills using drills and exercises

The National Fire Academy — A Jewel

The National Fire Academy (NFA) is our West Point and Harvard all rolled into one. It represents the best that the fire service has to offer. The library resources are the biggest and best available. Those who have not been there (and there are too many), think that most of their learning will come from the NFA courses. Those who have attended the NFA know that the course content is only a part of the learning process. The real value is in sharing experiences with firefighters from all over the nation in a first class learning environment. This experience is stimulating and broadens your view of the fire service. You will never be the same again. When you return to your own fire department, you will no longer have a local view of the fire service. You will return with a national view, increased knowledge, and a new network of friends.

THE NFA SYMBOL

Most of those who attend the NFA for the upper level chief officer courses are mid-level management: lieutenants, captains and junior chiefs. Very few chiefs who are the head of the department attend the NFA courses. Chiefs may not come because they believe that they are too busy, they already know all there is to know, or they may embarrass themselves with their lack of knowledge. This in unfortunate because the fire chiefs are the ones who can return home and make changes. **All officers should make every effort to attend the National Fire Academy.**

THE NATIONAL FIRE ACADEMY, EMMITSBURG, MARYLAND

SUMMARY

There is no such thing as a routine call. Take all calls seriously and be prepared because they all have the potential for injury or death.

When fire companies take responses seriously and are prepared, fireground operations will run more smoothly.

Remember that we play the game the way we practice and we set the stage for the big incidents at the little ones.

Training is vital, choose important subjects taught by good instructors and actively involve everyone in the learning process.

Make every effort to attend the National Fire Academy.

4

STANDARD
OPERATING PROCEDURES

"The success of any policy is measured by the catastrophes that do not occur." -Unknown author

The Night The Basement Fire Almost Ate Us

We started looking at SOPs in the Washington, D.C. Fire Department after we experienced several problem fires where things went badly. One night we responded to a reported fire on the top floor of a large three-story apartment house. Upon arrival, we saw smoke coming from the top floor windows, and we went into action. The top floor crews reported heavy smoke and no heat. They couldn't find the fire and began to think that it might be in the basement. It **was** in the basement, and fortunately it was extinguished without injury, but we were lucky. We had firefighters operating over an uncontrolled fire that we were not aware of. The potential for disaster was there. I don't know how many of you have ever seen firefighters bailing out of a building headfirst onto ladders; it's not something you soon forget.

**THE BASEMENT FIRE THAT
ALMOST ATE US**

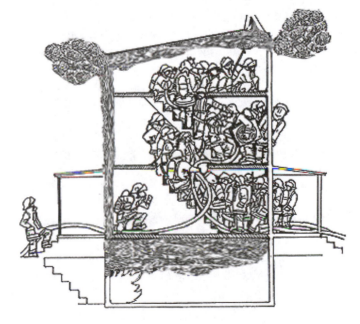

35

The Dangers Of The Fire Below

It is extremely dangerous to have crews operating over an unknown fire.

A United States Fire Administration report summarized the problems that one fire department had several years ago while operating above a fire without realizing it. As a result of this fire, one firefighter died and several were seriously injured. Excerpts from the report follow.

> On arrival there was light smoke showing from the top floor of a three-story abandoned warehouse. First arriving companies made their way up an un-enclosed stairway to the top floor where they found two small fires burning. They went to the window and dropped a rope intending to pull an attack line up to the third floor. The fire did not appear to be threatening at this point and the crews anticipated a quick and easy job of extinguishment.
>
> Very suddenly, the conditions on the top floor changed dramatically. Heavy smoke and flame rolled over the interior crews, forcing them to abandon their positions and retreat to the stairs. There were indications that an additional fire, possibly set on a lower floor had reached flashover and was rapidly engulfing the entire third floor. Of the eight firefighters that were on the third floor when the flashover occurred, five were trapped and went to windows to call for help. Two firefighters hanging from the window sills jumped before ladders could be raised for their rescue. One of these firefighters died and the other was seriously injured.

It is very dangerous for crews to be operating over an unknown fire, especially without hose lines. Standard operating procedures should be established to check the lower floors of fire buildings to prevent such tragedies.

Street Smart SOPs

After the basement fire almost ate us, we had two more fires that didn't go very well, and afterwards, when trying to figure out what went wrong, it became apparent that different people had different opinions on what the proper actions should have been. Like many departments, we have all kinds of rules and regulations about administration, disciplinary actions and uniform regulations. Our uniform regulations were so standardized that you were not permitted to wear a long sleeve shirt until October 1st, even if the temperature was below freezing. But we had no SOPs and few regulations to guide us in really important things—like handling our structure fires. It was basically a word of mouth operation.

Because of the lack of formal procedures, we started experimenting with some new procedures at the battalion-level. The changes came from the bottom up, not top down. In time, we made many changes that were eventually adopted as department-wide SOPs. These new fire and emergency SOPs have greatly improved our operations.

Check The Entire Building

Make certain there is no unknown fire below the fire attack crews.

Our SOP requires the second due engine company to check the basement and floors below the reported fire floor at all structure fires. In large buildings, the IC can assign other companies to assist in this task if necessary. The results are reported to Command. This SOP means a lot to the attack crews because they know that they are part of a system that prevents them from being caught by a fire under them. This procedure also helps to find the fire when its location is unknown. There is a tendency for firefighters to look where the smoke is, or where the caller reported the fire to be. Assigning a company to check each floor ensures that there are no hidden fires below the operating forces. This policy has improved our operations and increased our safety.

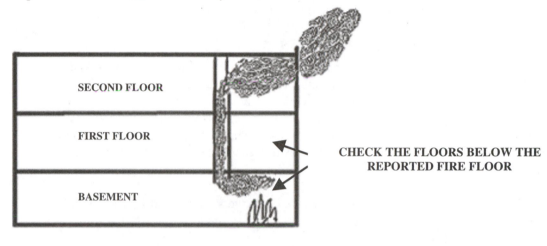

SECOND FLOOR

FIRST FLOOR

CHECK THE FLOORS BELOW THE REPORTED FIRE FLOOR

BASEMENT

Check the basement and floors below the fire at all structure fires.

The Cookie Cutter Approach
The vast majority of our incidents are replays of previous incidents.

Similar incidents should be handled by using the same basic approach. For example, most structure fires are in houses. While no two house fires are exactly alike, they are often similar, particularly in the same neighborhood. SOPs can outline what we normally expect each responding company to do at a structure fire.

By using SOPs to handle fires and emergencies, your operations will follow a pattern, and this results in consistent actions. The same things happen at each fire. There is always an accurate on-scene report, a water supply established, and attack and backup lines in use. The exposures are protected, the basement checked, searches performed, and salvage and overhauls handled. The point is, with SOPs all the bases are covered on every fire, so nothing is neglected. At each and every fire, the same things happen and the fireground becomes a predictable routine, improving your operations, accountability and safety. Without SOPs, it is more likely that things will go wrong, and when a few things go wrong there is often a domino effect—things get bad very fast.

Chief Dinosaur says, "No two fires are alike, so the SOP isn't worth a damn." Too often, this thinking is accepted in the fire service, but it is basically false. Think about it for a minute. Most of our fires and emergencies are similar incidents that occur over and over. By using

SOPs, every incident can be managed using the same basic approach. SOPs are also excellent guidelines for new officers, and all officers will have the confidence of knowing what is expected of them in emergency situations. Small departments with good SOPs often have better fireground operations than big departments with weak SOPs, because good SOPs make incidents run better.

SOPs are also valuable after an emergency during the post incident analysis. Comparing SOPs to actual operations help us to evaluate the actions taken.

SOPs Reduce Fireground Orders

Proper SOPs ensure that the correct procedures will happen with few orders given. When a working fire is verified, all first alarm companies follow the established SOPs. Each company notifies Command when they go to work as directed by the SOP. For example: *Engine 2 to Command, we are backing up the attack line.* In this way, accountability is maintained, and the IC is relieved of telling everybody what to do at every incident. This gives the IC time to think, to see the big picture, and to plan. The procedures make for smoother and more consistent operations.

Chief Dinosaur says, "We don't need no SOP, all they have to do is what I tell them." He *wants* to tell everybody what to do at every incident. The problem is that waiting for orders is slow, inefficient and lacks consistency.

 By using SOPs the average fireground needs very little command direction.

SOPs Avoid Consistency Problems
Without SOPs, every call is an adventure.

I overheard someone asking a chief how things went at a fire where that chief was not present. His answer was, "What chief was in charge?" meaning that incident management varies, depending upon who the IC is. Chief Smith operates this way, Assistant Chief Jones wants it another way.

Some departments have an A shift fire department, a B shift fire department and a C shift fire department. It is common in large departments to have areas of the city where operations are different than other parts of the city. The result is in-house jokes about how fires are fought in the "East End Fire Department" (referring to a section of the city.) In volunteer departments, who shows up can determine how well things go.

Standard operating procedures are critical for effective operation at emergency scenes. SOPs get us away from the "I'm here now and we'll do it my way," or "Let me tell you how we do it on this shift." Without SOPs, every call is an adventure, because there is no plan. SOPs provide a standard approach for each incident, reducing confusion and ugly surprises.

> We rolled up first due with a room off in the front of a house and started in with our line. As we were advancing, I saw the second line being stretched around to the rear. I thought to myself, "Oh no," and hoped for the best.

They took the line in through the kitchen window, charged it and fortunately never used it. Why they did it I don't have a clue, but that's what you get when you don't have SOPs.

Lieutenant Al Benton, Ladder 164. (Name changed to protect the innocent.)

SOPs And Lack Of Personnel

Make your SOPs flexible enough to compensate for the uncertain response levels that many fire departments face.

Chief Dinosaur says, "Hellfire, I just hope enough people show up to put the fire out. I don't have enough resources to do no SOPs." The chief has a point. Because of budget cuts, heavier EMS workloads and uncertain response levels, many departments have real problems gathering enough personnel to be able to follow SOPs.

The best way to deal with this problem is to make your SOPs flexible enough to compensate for the uncertain response levels that many fire departments face. For example, your SOPs can cover situations where only one company is available, what they should do and in what order, and how to operate a truck company with one person, two people, and so on. This balances the reality of fireground shortages with the need to operate within a SOP system. The wrong approach is to forget the SOPs for lack of people and just wing it.

Deviating From SOPs

"But the unexpected often happens," complains Lieutenant Sikorski. "What happens if the SOP says that my company should ventilate, but we see people hanging out of the windows?"

The answer is that SOPs must be flexible and allow deviation when necessary. Don't lose sight of the fact that SOPs are guidelines; company officers still have the ability and the responsibility to make decisions that may deviate from SOPs. When conditions make it necessary to deviate, company officers should report to Command (or Sector Officers) what they are doing, and what they are unable to do. This allows Command to be aware of what is happening and to make sure that all assignments are covered. An example of a deviation message follows:

Engine 3 to Command: We are not covering exposures, we are involved in rescue on side B of the fire building.

Deviation messages are critical. I remember a fire where an engine company assigned to backup the fire attack was sidetracked into medical care. They made a good decision because they came across several victims in the yard who were not breathing. I had no problem with what they did; I just wish they had told me about it. The original attack engine was delayed because they kept having to drag out the victims who were on the lawn. The result was that suppression was delayed. Had I known about the problems, I would have provided more help.

Whenever a company officer has told me that there was no time to report that the company was deviating from the SOP, I have told my Truck 3 story. At an apartment building fire, Truck 3 radioed me from the rear saying: *We are making a rescue back here, ventilation will be delayed.* It turned out they were making a gold medal rescue of a woman using their longest portable ladder and a pompier (or scaling ladder) to get her down from a fourth floor window. I've told company officers who didn't have time to call me, if Truck 3 was able to call me under those conditions, anyone can.

Enforcing SOPs

The wise IC can reinforce the SOPs by checking with companies to be sure that they are doing what they should be doing. SOPs must be enforced. If a company is not doing what they are assigned to do and freelancing, the IC or Sector Officers must follow up, ask why and let it be known that the SOPs **will** be followed. Too often, companies can do what they want to do and nobody says anything about it.

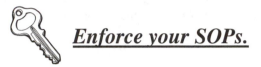 **_Enforce your SOPs._**

Writing And Reviewing SOPs
Good SOPs are basic, simple and easily understood by the troops.

SOPs should reflect the department's philosophy towards firefighter safety and consistency of operations. It should lay out responsibilities during emergency operations, and be consistent with laws and standards. SOPs must be written, simple and easily understood by the troops. They should include illustrations and diagrams to make them easier to understand. The more complicated the SOP, the more likely it will be ignored.

SOPs should be flexible and accepted by the officers and firefighters who will be using them. The best way to ensure acceptance is to include the officers and members in the development or revision processes. The advantage to using the participation and involvement approach is commitment. If the troops have a say in the changes, they may be more willing to change and offer good suggestions.

When writing or revising your SOPs, look over those of other fire departments to get ideas. Look at a mix of departments—small and large, career and volunteer to see the full range of philosophies and approaches. Obviously, the Podunk Fire Department can't adopt the New York City Fire Department's procedures, but they can look at the departmental philosophy and what the SOPs are trying to accomplish.

It is important to develop SOPs designed for your department. Don't just copy procedures from other departments, because you can easily wind up with procedures that don't fit your department. Look at what your experiences have been to learn what to expect in the future. For example, SOPs for basement fires are important in the northeastern part of the country. Your area may have problems with urban interface wildland fires. You need to know your problems and buildings to develop SOPs that apply to your area.

When writing or revising procedures, try to be open minded. Question current procedures. Why are we doing this? Is it reasonable? Do other departments do it differently? If so, why?

One department revising their SOPs discovered that their neighboring department normally used the second line stretched to backup the first line. That was different from their own procedures that required the second line to protect exposures. When they inquired why the neighbor department operated differently, they were told that the neighboring department considered the safety of their people ahead of exposures. This made sense. They revised procedures to require that the second line backs up the first line until personnel safety is assured; then they protect exposures. This change was to the department's credit. Many departmental egos don't permit adopting ideas from their neighbors. Doing so may be an admission that they've got something over on you. Contrast this with private industry where companies frequently borrow or steal successful ideas from their competitors to improve their own businesses.

Develop SOPs For Bad Days

Don't write only "sunny day" procedures.

It is easy to get too comfortable with SOPs that seem to work most of the time, but do not work for the worst incidents. It is when the unusual happens that the system is put to the test. SOPs should prepare personnel to expect the unexpected. Procedures need to cover the what ifs when things don't go as expected. For example, in one large department, a chief officer usually arrived on the fireground very quickly. Their SOPs were written so that the first arriving companies did not establish command, because it was assumed that a chief would arrive within a short time. Early one morning there was a serious commercial fire, and the chief was very late in arriving. All companies had self-committed by the time the chief arrived. The companies were doing inside fire fighting, but it should have been a defensive operation. The chief immediately pulled everyone out of the building, which collapsed after the last firefighter left the building. Following this fire, the procedures were revised to have the first arriving officer assume command.

Another fire department had a procedure that required incoming companies to report directly to the IC for orders, without staging. It worked quite well for a long time until they had a really major incident. The IC then became overwhelmed with incoming companies, and the system broke down. SOPs should be written for major incidents, not just everyday ones.

DESIGN SOPS FOR BAD DAYS
Courtesy of Fairfax County Fire and Rescue,
Fairfax, Virginia

Avoid Poor SOPs

Many fire departments have no SOPs, or they have SOPs that are so broad that they are almost meaningless. Poor procedures give SOPs a bad name, for example:

On structure fires, the first engine to arrive will take a position in the front of the structure, assume command and take appropriate action. All other fire companies that are responding will stage a block away in their line of approach.

What this SOP really says is: go to the fire and then figure out what to do.

41

A Sample Structure Fire SOP

The following SOP is designed to be used at most structure fires. It gives each company an assignment for typical interior fire fighting operations. This SOP uses management by exception. This means that the SOPs cover normal situations, so the IC needs to give directions only when circumstances are not routine.

The First Due Engine Company

Lay out to a hydrant, or other water supply, if possible

- Give a complete size-up report upon arrival

- Take a position in the front of the building, leaving room for a truck company

THEN

- The officer assumes command or passes it, depending on the circumstances

- The engine driver connects to standpipe and sprinkler connections where available, and charges the system at the proper pressure

- The engine crew advances an attack line or brings the standpipe hose into the building

Nothing Showing

The first due engine and truck companies enter the building prepared to fight fire. The other companies stage in their line of approach and monitor the fireground radio channel. If a working fire is verified, Command will normally direct all companies to follow the standard operating procedures as follows.

Working Fire

The first due engine company attacks the fire and is backed up by a second line. The backup line is normally the second line stretched after the initial attack line. **The backup line is very important** because the safety of the attack crew should not depend on a single hose line. These companies are in the Attack Sector.

The Second Due Engine Company

The second due engine company checks the rear of the building and the basement on **all** structure fires regardless of where the fire is or seems to be. The status of the rear and basement are reported to Command. Example:

Engine 2 to Command: Nothing showing in the rear and the basement is clear.

After verifying no fire in the basement, the second due engine company checks each floor as they go up, notifying Command of the status of the lower floors. Example:

Engine 2 to Command: Floors 1 and 2 are clear.

After checking the basement and lower floors, the second due engine company backs up the first due engine company, if there is not a backup line already in place. If there is a backup line, they protect exposures.

ENGINE COMPANY SOPS

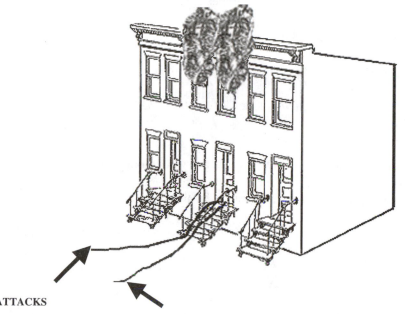

THE FIRST LINE ATTACKS

THE SECOND LINE BACKS UP THE ATTACK LINE

THE THIRD LINE COVERS EXPOSURES

The third hose line protects exposures, either interior or exterior. Interior exposures are usually on the floors above the fire. Before attempting to go above the fire, the exposure company should observe the fire location and size and the progress of the attack lines. If going up the stairway is not reasonable, and there are no exterior exposures, they notify Command of their availability.

Deviating from the SOP

When conditions make it necessary to deviate from the SOPs, company officers report to Command (or Sector Officers) what they are doing, and what they are unable to do.

Truck Company Operations

The truck company sizes up fire conditions and performs the following functions:

- Force entry

- Perform rescues

- Throw ladders and ventilate structure

- Check and ventilate the roof, if necessary

- Conduct primary and secondary searches

- Place lights in service and open up concealed spaces

- Shut off the utilities and perform salvage and overhaul

More detailed truck procedures are found in the Ladder Truck Operations chapter.

SUMMARY

SOPs should reflect department philosophy towards fireground operations and safety.

Good SOPs are basic, simple, and easily understood by the troops.

Using SOPs, operations follow a pattern, and similar incidents are handled consistently.

Proper SOPs ensure that the correct actions will be taken with few orders given. This relieves the IC from telling everybody what to do at every incident.

SOPs ensure that there is always an accurate on scene report and a water supply established. The attack line is backed up, and exposures are protected. The basement and floors below the fire are checked, searches are performed and reported, and the utilities are shut off. Using SOPs everything is covered at every fire, so nothing is neglected.

SOPs state what is expected, making them excellent guidelines for new officers.

Develop good SOPs by including the officers and members in the development process.

SOPs must be flexible and allow deviations when necessary.

SOPs are valuable during a post incident analysis to evaluate actions taken.

5

POSITIONING APPARATUS ON THE FIREGROUND

"The success of most fireground operations depends directly or indirectly on the effective use of fire apparatus" - Chief Alan V. Brunacini

When responding, the officer should check with the drivers to be sure that they both know the address and how they are due. If in doubt, they should verify the information with Dispatch. I remember a traffic cop stopping traffic for us twice at the same intersection as we tried to find our way to the fire.

Positioning apparatus on the fireground is one of the key elements to effective fireground operations. When the first alarm companies are positioned properly, it becomes easier to support their positions with incoming companies and to establish additional positions. Correct placement allows full use of all apparatus on the scene. When an apparatus is placed in a position that blocks streets or intersections, it can mess up the positioning of other companies. Have you seen an aerial ladder that can't be used when the truck can't get close enough to the building because it is blocked out of position? Have you ever seen apparatus parked on somebody's supply line? Use care so that your apparatus is not too close to the incident; your apparatus can make an expensive exposure.

Engine Companies

The first due engine should normally position in front of the fire building, just past the building entrance. All engine companies should leave room to allow truck companies to get a good position, and to pull ground ladders from the rear of trucks. Hoses will bend, ladders will not.

**ENGINES LEAVE ROOM FOR
THE TRUCKS**

Truck Companies

Trucks should be positioned for aerial ladder use. This will also allow easy access to tools, equipment and generators. Too often, trucks are simply parked out of the way, as though their main job was to transport the truck personnel to the fire. If the first due truck is not positioned right initially, you're going to be out of luck when all hell breaks lose. It will be almost impossible to reposition it because other companies will be in the way.

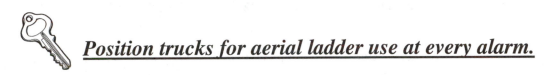

Position trucks for aerial ladder use at every alarm.

Recently I saw a house fire located at the end of a long driveway off a side road. The truck parked on the main road that led to the side road that led to the driveway. The truck was a couple of football fields away from the fire. It should have been at the end of the driveway, near the house, so it could be properly used. Sometimes truck drivers need to be aggressive in positioning their trucks. This may require jumping curbs or driving on lawns, driveways or parking lots.

**DO WHAT IT TAKES TO POSITION THE
TRUCK PROPERLY**
Courtesy of Fairfax County Fire and Rescue,
Fairfax, Virginia

46

When two truck companies are dispatched to an alarm, the truck company dispatched first usually takes a position in front of the building or area involved. The second truck company should cover the rear of the building, and, if possible, physically place the truck in the rear.

Medical Units

Medical units should position themselves close to the scene, and they should avoid getting blocked in or blocking subsequent arriving companies.

Command Vehicles

Because the chief's vehicle usually serves as the command post, it is very important to position it properly. The ideal position is the front of the building, with a view of the front and one side. When approaching the fireground, anticipate narrow streets and the direction from which most apparatus will be coming. To gain a good position, come in from the opposite side, use parking lots, driveways, sidewalks or vacant lots as necessary. This may take a little longer to do, but it will pay off in the long run.

Use care to avoid blocking other apparatus. The location of the command post should be announced over the fireground channel, so all the players know it.

A GOOD VIEW FROM THE COMMAND POST
Courtesy of Glen E. Ellman, Fort Worth Fire Department

There are plenty of real world examples of the problems caused by poor apparatus positioning. A nighttime fire in a commercial building in an urban department appeared to be a one line fire at the rear of the building. The fire attack was made through the front door, but the chief was parked near the rear of the building. The street was filled with fire apparatus, which would have required the chief to go around the block to get a good position in the front.

The fireground went to hell quickly, and the chief was not able to see what was happening. With no one outside the structure in a position to make informed command decisions, the situation quickly became a nightmare of lost firefighters and bad decisions. Poor communications compounded the problems.

Position the command vehicle in the front of the structure with a good view of the fireground.

At another fire, I arrived as the deputy chief at a working fire in the basement of a store. The battalion chief was positioned in front, only about 12 feet from the front wall of the fire building. I positioned my command post well back on the opposite side of the street. I called the BC on the radio, and he gave me a rosy assessment of the situation. It didn't look rosy from where I was sitting. I could see a lot of heavy black smoke boiling out of the rear of the building. I told the BC to come to the CP to change command. When he got there and looked back at the fire, his eyes got as big as saucers. He had lost the big picture from being right on top of the fire. It was a learning experience for him.

SUMMARY

Correct apparatus placement on the fireground allows full use of the apparatus and equipment on the scene.

The first due engine normally should assume a position in front of the building, allowing room for the truck company.

Trucks should be positioned to allow use of the aerial ladder if necessary.

Medical units should be positioned to avoid getting blocked in.

It is very important to position the chief's vehicle to allow a good view of the incident.

6

ENGINE OPERATIONS

"If you're going through hell, keep going" - Winston Churchill

This chapter focuses more on the philosophy of engine company operations than on procedures. Engine company work is not complicated. Everybody knows that engines "put the wet stuff on the red stuff." They usually go in through the door, find the seat of the fire and extinguish it. A quick knockdown eliminates most problems and improves safety. What follows are guidelines to help you standardize and improve the operations of your engine companies on the fireground.

Water Supply Guidelines

The first few minutes on the fireground are critical. Success or failure usually depends on controlling the fire quickly with an adequate water supply, so it is important to have water supply procedures. Many fire departments require the first arriving engine company to establish its own water supply by laying a supply line from a hydrant, if there is one within a reasonable distance. Laying a supply line going in is the best way to get a reliable water supply and allows use of both tank and hydrant water. It also allows the pump operator to be nearby, as well as tools or equipment that may be needed.

THE FIRST ARRIVING ENGINE LAYS A SUPPLY LINE

If staffing allows, a firefighter is left at the hydrant to charge the line when necessary. When a hose clamp is applied to the supply line after the engine stops at the fire, the hydrant person can charge the supply line and quickly rejoin his company. Some departments use an automatic hydrant valve that charges the line without the need for a hydrant person. When staffing is insufficient, the supply line is secured to the hydrant, and the firefighter re-boards to assist with the attack line. The supply line can be charged by the second arriving engine or other personnel. Valves can be used to allow an engine to hook up to the hydrant even after the original line has been charged.

Don't Pass A Hydrant

Many departments operate with the idea that the only time that you lay a supply line is when there is fire or smoke showing. If there is nothing showing, the first engine just drives up to the reported structure fire, and a few firefighters go in to look. The idea seems to be to avoid laying out supply lines needlessly. Let's call a time out here and examine this issue.

What do you stand to gain and what do you stand to lose by this approach? If there is a fire and you run out of water, what are the consequences? You could lose lives and the building itself. The only thing to be gained is saving a little extra work racking the hose if it's not needed. It comes down to convenience versus safety and efficiency.

ENSURE PROPER WATER SUPPLY

Without a supply line, you probably won't be able to supply both the attack and backup lines from the tank of the first arriving engine. This means that the backup line will be delayed, reducing protection for the initial attack crew. Laying a supply line mentally prepares you for a fire and increases the safety margin for both firefighters and the occupants of the structure. If you fail to lay a supply line because you don't think you have real fire, but find fire, the yelling and screaming begin. Fires can be tough when you are prepared, and even tougher when you are not.

Consider it from the point of view of the homeowner who calls the fire department. They expect us to come fully prepared. If **your** house had a fire that was not easily seen from the street, what would you want the first engine officer thinking about; saving time re-racking the hose, or saving your life and property?

If the first due engine fails to lay a line and relies on other companies for water supply, they risk running out of water. Some departments rely on the second engine to lay a supply line to the first engine. This can be a mistake because a delay in the arrival of the next engine is always possible and for a variety of reasons.

If the first due engine goes into the block without laying a supply line, the next engine often goes in thinking they will pick up the first engine's (non-existent) line. Then the truck comes and blocks everybody in. This problem can be avoided by doing the fundamentals. The first due engine lays a supply line and the second due supplies it. Engine companies advise where they laid lines from or that they were unable to lay a supply line.

At one fire, both responding engine companies passed hydrants and both used their tanks. The fire didn't go out, so one engine had to go back to the hydrant. How efficient do you think this operation was? What effect did lack of water supply have on fireground safety? How do you think the operation looked to the public?

The first due engine lays a supply line from a hydrant if possible.

Some of the traditionalists are probably saying, "What about all that hose that you stretched for nothing, all that wasted effort for those calls that were not working fires?" The answer is that no efforts are wasted; we are just doing our job by being prepared. Fireground operations usually go smoothly when having an adequate water supply is part of our routine. An old time chief once told me "Never be lazy on the fireground." He was right.

What's Wrong With Relying On The Tank
If the 500 gallons are gone and the fire isn't out, you are in serious trouble.

Most of us at one time or another, have been on a hose line that has lost water. It's scary and dangerous. When we fail to lay a supply line and rely on tank water, we risk running out of water.

Some fire departments in areas with hydrants choose to rely mainly on tank water. They are playing the odds. The odds are in their favor because 500 gallons of water can extinguish most structure fires. The problem is that if you play the odds often enough and win, you can develop a false sense of security. If the 500 gallons are gone and the fire isn't out, you are in serious trouble. Delays in getting water or running out of water, especially during the early stages of an incident, can result in firefighter injuries and fatalities. When operating from the tank, the pump operator must closely monitor the remaining water to give the inside crew advance warning if the water may run out.

**RUNNING OUT OF WATER
CAN BE FATAL**

Fast Or Slow Water
Because of the danger of running out of water, some fire departments delay charging attack lines with tank water until they have a guaranteed supply from a hydrant. The advantage of

51

this is that they should not run out of water once it is started. The disadvantage is that it is very slow and the fire can really take off while they are setting up. This means that the firefighters can be faced with a more dangerous fire because of the delayed attack. A quick attack with tank water while the water supply is being established is usually the best option.

Plan For Water Supply Problems

Water supply problems are not limited to rural areas. Even big cities have areas of poor or nonexistent water mains. Sometimes the city water supply is adequate for the ordinary fire, but is inadequate for large fires. Rural departments are sometimes better prepared for water supply problems than city departments because they know and expect that they will have water supply problems.

Big fires require big lines. Engine companies should lay large diameter lines or dual lines for major fire operations. Engine companies responding on greater alarms should lay additional supply lines to the first alarm engines that are close to the fire. Supplying first alarm engines saves time and fully utilizes the closer engines.

Areas Without Hydrants

Sometimes engines may have to depend on others for a water supply because of a lack of hydrants, excessive distances to hydrants, or reduced personnel. If there is no water source nearby, engines should lay lines that can be used when a water source becomes available. For example, the first engine should lay a supply line at a side road or driveway, so that the next arriving engine or tanker can quickly supply it.

I once witnessed a fire in a building that was back in the woods off a side road. The first engine failed to consider that a water relay would be needed. It was such a mess that the Ladies Auxiliary arrived and supplied cold drinks to the firefighters on the nozzle before they got water in their attack line. Not their finest hour.

Divide Your Water Supply For Safety

Don't put all your eggs in one basket. Fires like it when you take all the attack lines from one pumper because if something goes wrong (usually at the worst possible time), the fire can get away from you. The water supply should be divided among the engines for safe and reliable operations. Supply lines should also be divided to avoid loss of water supply. I witnessed a third alarm fire in a very large house where all the water came from one pumper on a hydrant. A breakdown of the pumper or the soft sleeve would instantly cause a waterless third alarm, resulting in a lot of yelling and scrambling for another water source.

Upgrade Your Water Capacities

Many fire departments have upgraded their attack lines to 1¾-inch lines which allow more water flow. Yet, they continue to operate off the standard 500- gallon tanks without laying supply lines. If you upgrade attack lines, consider increasing tank size as well. It is also a good idea to limit the number and size of lines that can be charged with tank water.

In areas without hydrants, a good option may be may be increasing the amount of water carried. Most engines have 500- gallon tanks, but 750, 1000 and 1500- gallon tanks should be considered. These larger tanks increase the chances of extinguishing the fire and allow more time to establish an alternative water supply.

Engine Company Basics

- The longer it takes to get water on the fire, the bigger the problem

- Advancing hose lines inside buildings often requires a big effort. Use multiple crews if necessary to get water on the fire as quickly as possible

- Firefighters should avoid running ahead of the hose line to the fire because it can be dangerous, and going ahead without helping slows down the line

- SOPs should require compliance with the two-in and two-out rule

- Engine companies should operate as one team, for safety and accountability

- The first due engine company should not go blindly to the reported fire floor. Use judgment: smoke will usually rise. Check lower floors as the line is advancing up stairways

- When opening doors in preparation for attack, engine companies should pause briefly at one side of the door to gauge conditions, and to avoid a possible blast of heat and fire coming from the opening

- Big fires require big lines

- Avoid directing streams into smoke except when there is very high heat and no visible fire

- Know the quickest way out

- Remember how you came in, and look for additional exits as you advance

- Avoid advancing hose lines up ladders and into windows for interior fire fighting

- Don't pass by fire; knock it down thoroughly before you advance

- Never let fire get behind you

- Do not allow hose lines to operate towards each other

- When hose lines are not advancing and conditions are getting worse, back out

- If a firefighter falls down a hole, a hose line directed down into the hole can save his/her life

When the engine team operates properly, the chances of a successful operation are greatly increased. One engine company in the right place at the right time is worth ten engines later. Extinguishment is basic to engine operations, but everybody doesn't always do it. An example was the lieutenant of the first due engine company at a high-rise fire who located the fire, then went up above the fire to the mechanical room to begin ventilation. Instead of attacking the fire, the first due engine company was playing with fans above the fire.

Arriving At The Scene — Don't Always Rely On What People Tell You

What people tell us when we arrive on the scene may not be accurate. You may be told that the fire is on the second floor because that is where the smoke is located. The fire might not be there. Don't be misled by well-meaning bystanders who may try to lead you around the building to show you the fire location. You'll often wind up in the backyard looking at fire and smoke coming out of a rear window, while your access is through the front door.

During a recent nursing home fire, an employee told firefighters that all patients had been removed, but firefighters later found several patients in rooms beginning to fill up with smoke. Be cautious and skeptical; listen to people, but use your judgment. Don't totally rely on what you are told. Sometimes people don't have to say anything, such as when you find old ladies standing in front of an apartment building at 3 a.m. in their nightgowns, covered with soot.

Use The Right Size Hand Line

It is not just the amount of fire that determines the line used, but also the layout of the building.

The conventional wisdom is "big fire, big line; small fire, small line," but it is more complicated than that. Generally, it's a mistake to use a 2½ -inch hose line for interior fire fighting at ordinary house fires. The 2½ -inch line is less maneuverable and usually provides more water than you need. It is not just the amount of fire that determines the line used, but also the layout of the building. House fires usually involve normal sized rooms, so one 1¾ -

inch hand line can usually handle the fire as you move from room to room extinguishing the fire. If, however, the same amount of fire is in one large undivided area like a store, you need at least one 2½-inch hose line. Learn to recognize situations where the 2½-inch hose line should be used.

SELECT THE RIGHT SIZE LINE

Another consideration is the personnel available. The reality today is that most fire departments are hard pressed to get an 1¾-inch hose line in service, let alone a 2½-inch hand line. Sometimes it may be necessary to combine engine companies to get enough personnel to be able to get a 2½-inch line into service.

 Recognize situations where the 2½-inch hand line should be used for fire attack.

Use Backup Hose Lines
The safety of the attack crew should not depend on a single hose line.

At a house fire, the first company ran their attack line to the fire. The second engine did not bring a backup line. This procedure was normal for this department. What wasn't normal was that the first due was unable to open the pipe because of a mechanical failure. The place was rocking with fire, and they had one useless hose line. It took a long time and much screaming and yelling to position the second line to begin the fire attack.

Most fires can be extinguished with one hose line; however it is important to backup the first company's attack line with a second line to the same position. The backup line is normally the second line stretched after the initial attack line. **The backup line is very important** because it provides protection for the initial attack line crew, increases knockdown power, and provides insurance against burst lines and other problems that may arise. Most injuries to firefighters occur when attacking the fire. The safety of the attack crew should not depend on a single hose line. After knockdown, the second line can be repositioned for exposure protection if necessary.

ATTACK LINE   BACKUP LINE

USING A BACKUP LINE

Some fire departments have problems with the placement of the second line. The second line is either sent to protect exposures above the fire or assigned to protect exterior exposures. While this may be necessary on occasion, remember that SOPs are designed for use in most situations. **When the first engine crew is inside attacking a fire, their safety must be your first priority,** and the best way to do this is with a backup line.

 ***Use backup lines for safety.***

Avoiding Flashover

What are the signs of flashover? There are a lot of them, but it is easier to avoid flashover than it is to look for the signs of it. How do you avoid flashover? First, ventilate and put water on the fire. If you are attacking the fire and releasing the heat and smoke, the risks of flashover are very small.

Engine Companies Rescue People With Hose Lines

The only time that an engine should delay suppression is for a life or death situation that obviously requires immediate action.

> In my fire department many think that rescue comes first no matter what and extinguishment comes later. I have seen fires roar out of control because the engines are searching to see if any people need to be rescued. Meanwhile nobody is doing a thing about the fire or water. — John Creighton, Engine 313.

Engine companies rescue people by putting out the fire. Rapid extinguishment eliminates most problems, including the need for rescue. Our first priority is rescue, and that leads to a common question. Under what circumstances should rescue delay fire attack? The only time that an engine should delay suppression is for a life or death situation that obviously requires immediate action. This does not mean assisting people who might be in trouble, or helping people who are self evacuating. It means rescuing people who are clearly in immediate danger, such as people hanging from an upper story window. This is when rescue comes first, assuming that there are insufficient resources to do both rescue and suppression at the same time.

 Putting out the fire eliminates most rescue problems.

Direction Of Attack

Most tactics books include a dozen or more hose line principles such as attack from the unburned side, protect means of egress, and so on. Most of the principles are sound, but they can be difficult to carry out on the fireground. For example, when a fire is located on the first floor toward the front of a house, it is typically extinguished by attacking through the front door, thus violating the attack-from-the-unburned-side rule. But attacking the fire from the unburned side often results in delaying the initial attack. First, you often have to go around the house and climb a fence. Then you go through the clothesline, fight the dog, force the door and start crawling through smoke-filled hallways to find the fire. All the while, the fire continues to get worse. Then when you finally attack the fire, you have to worry about finding your way back out if things get bad.

It is much easier to just go in through the front door and put the fire out. It may violate somebody's principle, but it's easy, it's safer, and it works. You don't drive fire that you put out. Usually, most fires involve a room or two that can be rapidly extinguished by an engine company crew.

Normally the rears of commercial buildings are locked up like Fort Knox after business hours. The fronts of these buildings are often glass. Usually it is a mistake to attack these fires from the rear. Besides, we do most of our fire fighting from the front and most companies are usually positioned there.

An example of problems with rear attacks was shown to me when I was visiting a fire station in another city. One of the local buffs showed up with a video of one of their recent fires. The fire was in a two-story automotive parts store without exposures.

At the start of the videotape, there was light smoke coming from the front of the building. It looked as if an attack and backup line from a pair of aggressive engine companies would control the fire. That never happened. Nothing much ever happened except for master streams much later in the operation. I had never seen anything like it. It almost looked as if they wanted to lose the place. My hosts realized that it was a major screw-up. The explanation they volunteered (I was too embarrassed for them to ask what went wrong) was that the plan was to attack from the rear, but communications were poor, and they never did get rear access. The building burned to the ground.

It occasionally happens that companies in the front have difficulty gaining entrance. Under these circumstances, other companies should try to gain entry from the rear. No rear attack should be started without a green light from the IC who coordinates the operation. The IC should give the front companies a reasonable time to get in and knockdown the fire. If the companies in front are unable to get in and make progress, back them out and attack from the rear.

Attack directions are not carved in stone. If there are problems with attacking the fire from one direction, command should switch to another.

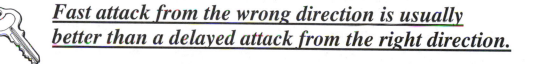

Fast attack from the wrong direction is usually better than a delayed attack from the right direction.

Bending The Interior Attack Rule
Beware of the terms "never" and "always" as there are few absolutes in fire fighting.

Interior attack is usually the best way to go, but not always. I remember a fire in our downtown district years ago. The fire was showing out of a second floor window and appeared to be a one-room fire. The first floor was commercial and locked with a security gate. It was taking forever to gain entrance, and an engine company was used to knock the fire down from the outside, while the forcible entry was underway. Being young and dumb, I thought that we had done the wrong thing by breaking the "Always go inside, never put water through the window rule." A wise old lieutenant asked me "What were we supposed to do? Let the place burn down according to the rule book?" His point was that, while rules are necessary, we need to be practical and flexible in the fire fighting business.

Years later, I bent the same rule as a battalion chief. The fire was located on the second floor of a very large high school. Fire was coming out one window, and it looked as if we had part of a room involved. It was a Sunday morning, so the building was empty and heavily secured, and it was a very long hose stretch. A ladder was raised to the room, and an engine company was standing by with a charged line. They were eager to knock the fire down, and I was too. I checked with the engines to see if they were getting ready to hit it from the inside. They were not even close to the fire yet. I told them that we were going to hit it from the outside. It was a simple job to knock it down, and it was mopped up from the inside. I soon heard the same criticism that I had made year's ago about the "always hit it from the inside rule." I gave the same answer that was given me: sometimes the rules need to be bent. Keep in mind that in both fires, the buildings were unoccupied, the interior attack was seriously delayed and the outside attacks were coordinated through the IC.

Hose Streams From The Outside

Occasionally an interior attack is hampered and endangered when some well-meaning but ignorant troops direct a hand line in from the outside on top of the inside forces. At one fire the radio screamed: *Attic to Command, have the outside line into the attic shut down!* Command replied: *They are just helping you a little.* That kind of help we can do without. We tend to focus on the problems with outside lines from master streams, but often the problem is the stray outside hand line "helping" extinguish the fire.

NEVER USE OUTSIDE LINES ON INSIDE TROOPS

 Using outside steams on inside troops is like being bombed by your own air force.

After Fires Are Knocked Down

When the fire is knocked down, the Attack Sector should:

- Notify Command

- Keep SCBA on

- Vent the area

- Quickly check and report on the overhead; pull ceilings and check above

- Always check the attic; fires anywhere in a building can extend there

SUMMARY

The first few minutes on the fireground are critical because success or failure usually depends on controlling the fire quickly with an adequate water supply.

The best way to get a reliable water supply is for the first arriving engine to establish its own water supply by laying a supply line from a hydrant.

When you rely solely on other companies for water supply, you increase the risk of running out of water.

Avoid having all the supply lines feeding one pumper or having all the attack lines come from one pumper. It is safer to divide the water supply among engines.

Big fires require big lines. Engine companies should lay large diameter or dual lines for major fire operations.

Engine company work is not complicated: go in through the door, find the seat of the fire and extinguish it. Usually this simple approach is the best, but if there are problems with attacking the fire from one direction, switch over to another.

Engine companies should operate as one team, for safety and accountability.

The backup line is very important because it provides protection for the initial attack line crew, increases knockdown power and provides insurance against problems that may arise. When the first engine crew is inside attacking a fire, their safety must be your first priority.

Engine companies rescue people by putting out the fire. Rapid extinguishment of the fire eliminates most problems, including the need for rescue.

Engines need to avoid the mind set that what is normally done at most fires is needed at all fires. Remain flexible and do not get locked into "one size fits all thinking."

7

Ladder Truck Operations

"We don't get any of the glory, but we are a critical part of the operation. We are the ones that make it all happen. Our venting and opening up lets the engines do their job and get out alive." - Anonymous truck officer

This chapter is much more detailed than the chapter on engine operations because truck company operations are much more complex, and less well understood. The truth is that there are not many people who really understand truck work. Even in large fire departments, really good truck companies can be hard to find; many are mediocre, and some are lousy. This same ratio is probably true in smaller fire departments.

Chief Dinosaur says, "Trucks don't have sufficient staffing. What do you expect?" The answer is that no trucks have enough staffing to do what is needed right away. An understaffed truck should do all the right things, but more slowly.

This chapter provides guidelines to help standardize and improve the operations of your truck companies on the fireground.

WHY ARE GOOD TRUCK COMPANIES HARD TO FIND

Some departments are engine company happy; they fight fires with the engine-only approach. Firefighters are usually more comfortable doing engine work than truck work. In engine companies, the mission is clear, and there is an immediate payoff when the fire is extinguished. Truck work is more complicated and diverse. Truck firefighters often feel that truck work is for thinking firefighters, while engine work is basic and simple. Engine people often wonder why anyone would want to run around with a hook or a fan, when they could be putting out the fire. Both engine and truck work are needed, and the work of one complements the other. Everyone must recognize the importance of truck work. Many departments cross-train firefighters for both engine and truck work. While this may sometimes be necessary, permanent crews are desirable to improve truck company operations.

Poor Truck Work

A fire in a high-rise building was captured on video. It began with fire showing from several windows on the fifth floor of a twelve-story apartment building. A truck company raised their aerial ladder to the balcony next door to the apartment on fire, and two firefighters were standing on the turntable. No one climbed the ladder. The aerial was then removed from the balcony and positioned in the air for a while, and then moved back to the balcony.

Next, a firefighter climbed the aerial ladder to break out the glass doors of the apartment next door to the apartment on fire. The ladder then moved away leaving the firefighter working alone on the balcony, unable to retreat back down the ladder if necessary. Later the ladder was back on the same balcony with two truck firefighters again standing on the turntable of the truck. At this point, the fire was not controlled, and the ladder was not being used. I don't know why the firefighters were standing there, and I doubt if they did either. The fire was in several rooms in a large occupied apartment building, so obviously there was a need for inside truck work.

Why did this truck company perform so poorly? It was because the truck company crew was inexperienced and poorly trained. Sound familiar?

Don't Let Your Trucks Become Engines

Truckies on hose lines are a common problem in many fire departments.

It's basic—engines do engine work and trucks do truck work. If truck firefighters are operating hose lines, something is seriously wrong. Truck company members do not belong on hose lines. "I've heard time after time," one fire officer says, "We couldn't do truck work or search because there was too much fire," and this was at a simple room fire.

There is no truck company anywhere in the world that arrives with more people than are needed to perform the necessary truck work at a working fire. Nevertheless, truckies on hose lines are a common problem in many fire departments. Why? The reason is mostly a lack of understanding, training and discipline. We need to realize that failure to do truck work places engine crews and victims in greater peril.

KEEP TRUCKIES OFF HOSE LINES
Courtesy of Fairfax County Fire and Rescue, Fairfax, Virginia

"But We Don't Have A Ladder Truck"

The problem often lies in the mind-set, of the firefighters, not in their lack of equipment.

When fire departments do not have truck companies available to respond, engine company personnel are used to do truck work. It is vital that truck work be done whether you have a truck company or not. Engines can carry most of the basic truck equipment. The problem is often in the mind-set of the firefighters, not the lack of equipment. Engine people tend to think engine work, and it can be difficult to get the truck work done. Firefighters should be trained and encouraged to develop truck work skills. Those showing truck talent should be used in truck work whenever possible. The best way to get the truck work done is to assign a truck crew as soon as possible.

IMPROVING YOUR TRUCK WORK

To begin improvement, understand the value and importance of truck work. Next, develop truck company SOPs that improve truck operations. If truck company procedures are not clear, you cannot blame the troops for poor performance. Training, pre-planning, group discussions, drills, and post incident analysis are vital to successful truck operations. Too often, an excellent truck person gets promoted or transfers to an engine company where their talents are wasted. While both engines and trucks are important, talented truck people are rare, and their talents should be cultivated and not wasted in an assignment that could be easily performed by other people.

TRUCKIES DO TRUCK WORK

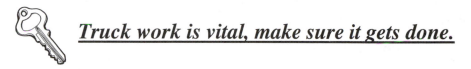

Truck work is vital, make sure it gets done.

Truck Teamwork

Truck company teams enter the building together, work together, and leave together.

At most fires, truck company officers should not have to give detailed orders. Truck personnel (or those assigned to do truck work) should know in advance what is expected of them. They should think as individuals and work as a team and focus on the most urgent concerns first.

Operate with two teams if possible–an outside and an inside team. This requires at least four people, including the officer. The truck officer operates with either team as the situation requires. When there is insufficient staffing for two teams, use one team.

Operating with one team

- Size-up fire conditions and perform obvious rescues

- Raise portable ladders

- Vent from the outside

- Check the roof and take off roof hatches or skylights and give Command a roof status report

- Go inside as a team with full gear, SCBA and tools

- Ventilate

- Conduct primary and secondary searches for victims

- Report status of both searches to Command

- Place lights in service

- Open up concealed spaces to expose hidden fire

- Perform salvage and overhaul

Operating with two teams

- Both teams size-up fire conditions and perform obvious rescues.

The outside team

- Raise portable ladders

- Vent from the outside

- Check the rear and sides of the building

- Check the roof and take off roof hatches or skylights and give Command a roof status report

- Control utilities

- Join the inside team when the outside work is completed

The inside team

- Force entry if necessary

- Go inside as a team with full gear, SCBA and tools

- Ventilate

- Conduct primary and secondary searches for victims

- Report status of both searches to Command

- Place lights in service

- Open up concealed spaces to expose hidden fire

- Perform salvage and overhaul

Divide The Truck Work

If a second truck company is available, the truck work should be divided. A good method is to assign the first truck to the fire floor, and the second truck to the floors above the fire. If you don't have a truck or a second truck, engine companies should be assigned these duties.

Truck companies should communicate with each other to coordinate:

- Rescue operations

- Roof ventilation

- Evacuation, particularly the best method or route

- Ventilation
  - type of ventilation being used
  - direction of smoke movement

Truck Personnel Accountability

Officers must keep track of their firefighters.

It is important that truck teams enter the building together, work together, and leave together as a team, so that no member operates alone. Truck officers usually cannot directly supervise all their people on the fireground. Officers must keep track of where each member is, what they are doing, and when they should be expected to report back. This is in addition to other accountability systems that may be in use. If members do not appear or report when expected, the truck officer must initiate a check for their safety. Truck members are also responsible for their own safety and for staying in touch with their truck officer. Making this happen requires much training so that the troops cooperate and buy into the accountability system for their own safety. One California fire department has signs on their fire station mirrors that read: YOU ARE LOOKING AT THE PERSON WHO IS RESPONSIBLE FOR YOUR SAFETY.

Portable Ladder Placement

Proper placement of ground ladders is important for fireground rescue and ventilation. Some fire departments only put up ground ladders when there is an obvious need. We should be putting up ground ladders whenever we can, for the safety of our own. The ladders provide a second way out. A ground ladder in the right place and the right time could save a life.

Ground ladders are used for:

- Rescue
- Venting
- Exit for firefighters (When firefighters have entered a building by way of a ladder; don't remove the ladder until all members have exited.)

Put up portable ladders.

**PUT UP GROUND LADDERS ROUTINELY
FOR FIREFIGHTER SAFETY**
Courtesy of Allan Etter, DC Fire/EMS

Ladders should be raised to:

- The fire floor
- The floor(s) above the fire
- The roof, if needed

When raising ladders use care to avoid overhead electrical wires.

PUT UP LOT'S OF GROUND LADDERS
Courtesy of Allan Etter, DC Fire/EMS

Aerial Ladder Placement

Truck companies usually drill with the aerial ladder at the fire station or at a fire training building. They use the same structure over and over again. If that's where the fire is, no problem. But the real world has parked cars, apparatus, trees, wires, light poles, long lawns and other problems. Only experience will make you proficient. What better way to be prepared than to use the aerial ladder at small fires just for practice. Chief Dinosaur says, "The aerial ladder is usually not necessary; only use it when y ou need it."Dinosaur's right that the aerial ladder isn't needed very often. The problem is that if you only use it when you desperately need an aerial ladder, it probably won't be in the right position, and the operator won't be very good at maneuvering the ladder through the obstacles. When the ladder is used routinely, the truck will be in the right place, and the operators will be sharp.

USE YOUR AERIAL LADDER ROUTINELY
Courtesy of District Chief Chris E. Mickal, New Orleans Fire Department

FORCIBLE ENTRY

A mid-sized career fire department was dispatched for smoke in a nursing home. They saw light smoke in a store room through the window of a locked door. The fire companies waited around for the key that was supposed to be coming from the front desk. The fire developed into a multiple alarm requiring a partial evacuation of the nursing home. Their reluctance to force the door could have resulted in the deaths of nursing home patients.

A firefighter in another department told me that he forced the door at an obvious house fire. His chief was waiting for a neighbor who went to get a key and was not happy about the forced door.

What causes these screw-ups? The reasons are probably a fear of excess damage and the need for a reality transplant. It's okay to avoid unnecessary damage, but some people get silly about it. Trying too hard to avoid damage during fires can increase life and property losses, not reduce them. A old 'bars and irons man" once told me, 'I got the key right here bo ss,"as he put the halligan bar into action to force a door.

VENTILATION

Fire fighting inside poorly ventilated structures is dangerous and glass is cheap.

Ventilation is basic and critical; however, some fire departments are still not ventilating properly. We burn down too many buildings and hurt too many people from lack of ventilation. **Vent early and vent often.** Don't screw around at working fires; bust out the windows using ladders and tools. Most rescues are accomplished by venting the windows from the outside, attacking the fire and searching. Releasing smoke and heat increases visibility, and improves the chances of survival for both victims and firefighters. Proper ventilation allows firefighters to be more effective and reduces or eliminates the chances of flashover and backdraft.

AN UNVENTED BUILDING IS A DEATHTRAP FOR THE INSIDE FIREFIGHTERS.

The windows at a recent store fire were covered over with plywood. Interior crews were inside trying to find and extinguish the fire. Visibility inside was poor, and heat levels were increasing. The interior crews never relayed the ventilation problem to Command. The truck company never even started opening up the windows. It turned out that two of the truck people were on the roof and the other two were inside on a hose line. A later arriving truck started to open the windows, but it was fifteen minutes before they were vented. Command never asked the Interior Sector about smoke conditions and never checked with the truck to see what they were doing. This fire went downhill fast. Sudden heat buildup and zero visibility drove the firefighters out of the building in a forced and scary evacuation. Interior crews rolled out the front door so hot that they were smoking. **At fires that turn sour, lack of ventilation is almost always a major contributing factor.**

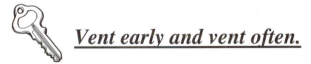 *Vent early and vent often.*

One morning a truck company with three people arrived at a working house fire. One firefighter was supposed to ventilate and ladder the outside while the other two went inside to do truck work. The outside firefighter couldn't do much alone. Besides, he wanted to get inside because he thought that's where the action was. He was right. There was plenty of action inside, mostly because companies were stumbling around in high heat and no visibility because no one had vented the windows. Eventually, another company vented and things dramatically improved.

Another example of the results of poor ventilation occurred recently, during an after business hours fire in a commercial building. The place was in a high crime area so forcible entry was formidable and time consuming. The truck had four people: one was assigned to forcible

68

entry; the other three went to the roof to begin ventilation. It took forever to gain entry into the building, and the premature roof ventilation hurt the operation.

Avoid using poor truck SOPs, like the one from a large urban department with four person crews. They send two to the roof, and the other two go inside. Needless to say, there is very little portable ladder work or outside ventilation done in this department.

Ventilation is one of the most important things that trucks accomplish. Truck companies should not be in such a hurry that they skip ventilation to rush into to a smoky hot box building. If you don't have enough people for both an inside and an outside crew, stay outside until you have the place vented. Trucks should ladder and vent from the outside before they go inside at working fires, unless they have enough personnel for both inside and outside teams.

Trucks operating with one team should ladder and vent from the outside before they go inside at working fires.

Glass is cheap and fire fighting inside poorly ventilated structures is dangerous. When in doubt break it out, taking out all the glass in a window. Quickly clear away drapes, curtains and screens and other obstructions to speed the venting. Window venting saves firefighter and civilian lives by reducing smoke and carbon monoxide levels, improving visibility, reducing heat, and giving firefighters a second way out. Beware of officers that are into saving glass because their priorities are wrong.

I once read in a tactic's book to vent windows by opening the bottom 1/3 of the way up and the top 2/3 down. This had to be written by some pencil pusher who never went to a fire. At serious working fires, we want the windows to be **gone**, totally open, not just played around with. Some books get fancy about ventilation, such as venting to draw fire away from trapped occupants. It sounds great and under the right circumstances it would probably work. The problem is, it is very difficult to know for sure both the exact location of the trapped occupants and the location of the fire. It is usually better not to be too fancy and just do the basic ventilation. The name of the game is to open up quickly. Slam, bang, and it's vented.

Fast Window Ventilation

Proper venting is often the difference between success and failure.

Using ground ladders to break out the windows is often the fastest and safest way to ventilate a building. Ladders are great for taking out windows—even in a one-story ranch house. First floor windows are punched out by smashing the end of the ladder through the window. Second and third floor windows are taken out by raising the ladder to the height of the window and then forcibly dropping the ladder into the window. Face shields and gloves should be used and heads should be down at impact because of flying glass. It may take several drops to clear out the window. The ladder is then slid or flipped across the side of the building to the next window. Truck companies can break out a lot of windows this way using the bulk of the ladder as a tool and avoiding the ladder climbing hazards of falls or fires jumping out at the firefighters. It's not fancy, but it is very fast and effective. Beware of contact with electrical wires that may occur if you are working fast in smoky conditions and not paying attention to the ladder tip. Occasionally, it may be necessary to go up the ladder to break out the windows with tools if this method fails.

VENTING WINDOWS WITH A LADDER

**VENTING WINDOWS BY CRASHING THE LADDER INTO THE WINDOW
AND FLIPPING THE LADDER TO THE NEXT WINDOW**

Energy Efficient Windows

Wood-sash single pane windows tend to let smoke out and are easily broken, either by heat or the fire department. Double-paned energy efficient windows cause problems by keeping the smoke in, delaying detection, and increasing heat and smoke conditions. It can be difficult to recognize the extent or location of the fire upon arrival. Look for and identify these types of windows quickly, report them to Command and take them out quickly. These windows are difficult to break out, requiring bars, axes and a lot of effort to make sure that both panels are smashed.

Hose Line Ventilation

Once the fire is knocked down, the hose line is already there, and it may take time to set up a fan operation. Venting with a fog stream out the window of the fire room works great. The nozzle is held a few feet *inside* the window, and a fog pattern is adjusted to fill most of the window space. It is fast and effective. Try it.

WINDOW

3 TO 4 FEET

VENTING WITH A FOG STREAM

Venting Windows with an Aerial Ladder

I once taught a class where a student showed a video he had taken of a New York City truck company using an aerial ladder to punch out some fifth floor windows to ventilate a building. Some departments do this routinely, while others recoil in horror at the thought. The reaction of the class was mostly disbelief and fears that the ladder may become damaged. Most aerial ladders can be used to punch out windows, but not all. Be guided by the manufacturers' instructions on the limitations of their ladders. Punching out windows can be a fast, safe and easy way to ventilate. Remember that rapid ventilation improves the life safety of both occupants and firefighters, and keeping the end of the aerial ladder pretty isn't a high priority at fires.

PUNCHING OUT WINDOWS WITH AN AERIAL LADDER

I was once a battalion chief in charge of a fire on the fourth floor of an office building. Smoke conditions were heavy on the fire floor, and I was getting repeated, urgent calls from the interior for more ventilation. I told a truck company in the front to start punching out the windows after clearing the area below. Just before they started on the windows my boss, the deputy chief arrived. I explained the situation to him and what we were about to do. Just then the tip of the aerial ladder crashed through the window, and no smoke came out. The deputy looked at me, looked at the window and then at me again. The truck operator said, "Do you want me to keep doing this?" I told him to keep going, knowing that there were severe smoke conditions and thinking that maybe we had opened the wrong window. We took out about five or six more windows, and still no smoke. I wasn't sure what was going on, but it seemed we weren't accomplishing anything by punching out windows. Very soon, the Interior Sector began reporting improved smoke conditions and soon the fire was controlled. It turned out that the punched out windows were creating an inlet for ventilating the smoke that was going out the back. The tactic worked, it just didn't work the way that I thought it would.

Ventilation Timing

A great deal has been written about ventilation timing. Sometimes it appears that ventilation timing is so critical that we need stopwatches to pull it off successfully. It's not that complicated. When the attack crews are ready to go, the vent crews start venting. Don't vent until hose lines are ready to advance. The only exception to this rule is the rare occasion when you have a solid report of a victim in a specific room and there is no engine company on the scene. In that case, the vent crew should vent for life by breaking the window of the room from the outside, before beginning a quick search.

Once hose lines are in action, open up all you can. Most ventilation problems are not in timing, but in the failure to open up quickly and thoroughly.

Positive Pressure Ventilation (PPV)

Positive pressure venting can be fast, safe, and efficient.

Positive pressure venting uses gasoline-powered blowers, forcing fresh air into a structure, expelling smoke and other gases. There are several advantages to positive pressure venting:

- Speed
- Efficiency
- Personnel safety
- Convenience of doors and halls not blocked by fans. (How many times have you hit your head on a fan in a doorway?)

The PPV blower is set up four to eight feet from the outside entrance on the unburned side, with the cone of air from the blower covering the opening.

POSITIVE PRESSURE VENTILATION

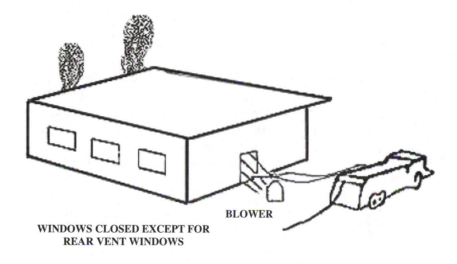

WINDOWS CLOSED EXCEPT FOR
REAR VENT WINDOWS

BLOWER

PPV During Fire Attack

PPV can work very effectively in many interior attack situations by pushing heat and smoke ahead of the advancing hose lines. Using PPV properly during fire attack requires extensive hands-on training. It is important that command officers be aware of PPV's advantages and limitations to know when to use it properly. To use PPV, you have to know all the following:

- The exact location of the fire and any victims
- The status of all openings, and where heat and smoke will vent out
- The attack team, ventilation team, and Command are ready for PPV to be started.

The ventilation team makes an exhaust opening if necessary. The attack crews must be ready to go as soon as PPV is started. As they advance, the heat and smoke is pushed ahead by the blower, which should improve the speed, efficiency and safety of the attack teams. To increase the safety of interior crews if one blower fails, two blowers can be placed in series. To do this, place the first blower two feet from the opening so that all air from the blower blows in. (Do not try to seal the opening with this blower.) The second blower is placed six feet behind the first blower, with the cone of air from this blower covering the opening.

Real World PPV

PPV is not a cure all for ventilation, it is only an option. While positive pressure ventilation can work very well in some situations, do not use PPV during a fire attack if:

- The location of the fire is unknown

- The location of possible trapped victims is unknown

- The path that the smoke will vent out is uncertain

Using PPV under these circumstances may spread the fire and endanger any victims or search teams between the fire and the vent openings.

When PPV cannot be safely used, quickly start breaking out windows from the outside. Firefighters inside an unventilated building are at risk, and the more ventilation the better.

When the fire is out and you just want to get rid of the smoke, positive pressure is very effective. Ventilate one room at a time by opening doors and windows (after removing screens.) Electric fans can be useful inside structures because they do not produce carbon monoxide, and can be used in conjunction with gasoline-powered blowers. When operating PPV blowers inside, monitor carbon monoxide readings to ensure that the CO levels do not get too high. When readings get high, shut off the blowers and use electric fans until safe CO levels are reached. When the room is clear, close the openings and continue venting the next room or area.

PPV In Large Buildings

When venting large buildings, begin PPV in the area of the fire. Then, divide the remaining areas and vent each in turn, starting with the areas having the heaviest smoke conditions. Proper use of the building's ventilation system fans can help the operation.

In multi-story buildings, the stair shaft can be used to achieve PPV to remove smoke from the building. Pressurize one or more of the stairwells, using caution to avoid venting through stairways that people are using to exit the building. Pre-plans are helpful to show which stairways have doors, hatches, or skylights on their topside.

Using PPV to remove smoke from a floor usually begins by placing blowers covering the entrance to the stairway on the fire floor. Open the roof opening over the stairwell and keep the stairwell doors to the other floors closed. It is not necessary to move blowers to vent the remainder of the building. When the fire floor is clear, close the door and open the door to the floor above the fire to vent that floor. After each floor is vented, close the stairway door. Vent each floor in turn, using this method. It is important for one person to be in charge of the PPV operation so that it is properly coordinated.

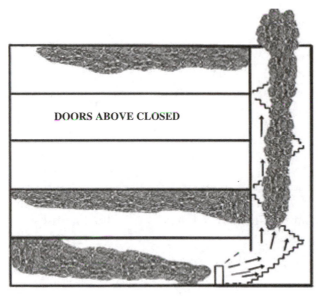

DOORS ABOVE CLOSED

PPV IN LARGE BUILDINGS

BLOWER

SEARCH AND RESCUE

We don't have trapped victims every day. Because of the infrequency of rescues, we sometimes imagine when they will occur. We think about the 3 a.m. house fire, with cars in the driveway, and fire and smoke showing. No occupants are outside, and the neighbors are screaming that they are still inside. Trouble is that it often doesn't happen this way. Sometimes there are victims when we least expect them, in the afternoon, in abandoned houses, or in vacant stores. In one case, a homeless person died in a fire in a "one-hole" portable construction toilet, and was overlooked by the fire company that extinguished the fire. Why? Because it was 4 a.m. and the toilet was located in an isolated construction area, so the fire company assumed that it was vacant. They couldn't have saved the victim, but they could have saved themselves a lot of embarrassment.

One afternoon, we had a working fire in a very large residential garage. I was with the first due engine, and after we had the fire knocked down, I started a primary search. I took one step and stepped on something squishy. "Oh God," I thought, "I stepped on a body." Shining my light down through the smoke I saw what appeared to be a zebra. You don't find zebras during primary searches in Washington D.C., so I continued my search for victims. The secondary search confirmed that I had indeed found a zebra. Turned out that the garage was being used for a taxidermy business. Though I found a zebra, it could have been a victim—assume nothing.

ASSUME NOTHING

Search Procedures

Our first priority is to save lives, and our procedures should reflect this. The basics of search and rescue are to vent, enter, and search. A planned search should be made at every fire, not just when you suspect that there are victims. People can be trapped at all structure fires—don't believe that everybody is out because:

- The building is abandoned, or

- The building is locked up, or

- No one is reported trapped

Your fireground rescue procedures should indicate what search actions should be taken in various rescue situations. Too often a specific team isn't assigned to search. "In my department, everyone wants a hose line; organized searches usually are not done," a disgusted firefighter once told me. He also said that truck companies often pull lines to protect their search, and end up fighting the fire and forgetting the search. **Time out!** We say that our first priority is to save lives, but our procedures sometimes don't reflect this. Everybody talks search, but do we really do it?

Many fire departments really don't search much. I watched a television interview of a firefighter who was credited with rescuing a fire victim. Asked to describe the rescue he said, "I was walking through the bedroom with a fan and I tripped over the victim lying on the floor." We tend to focus on obvious rescues like occupants at the windows or people who need help finding their way out, sometimes neglecting to search for unconscious or hidden victims.

Search is normally a truck function, but search assignments are often unclear. We don't wonder if there will be a pump operator at a working fire. We realize that the pumps are a vital function, so somebody is assigned to them at every fire. Since saving lives is our number one priority, how can we fail to have a search team? If you haven't assigned search teams, it means you only talk search. If we really consider search to be important, a search team will be designated for search and rescue.

Assign a search team at structure fires.

Searches should start in the right places. I remember one fire located on the first floor of a two-story house where search teams started the search on the floor above the fire, doing both primary and secondary searches before any searches were started on the fire floor. I have heard about a fire where, unbelievably, the search team started the search on the floor *below* the fire.

When you have credible reports of people trapped, get specific locations about where the victims are likely to be found. Without specific information, search the fire floor first because most victims are found there. The next most likely place is the floor above the fire and then the top floor. Bedrooms should be given a high priority for search because many occupants are found there. If heat and smoke prevent entrance to an area, probe near doors and windows, because victims are often found near exits. When many people are trapped, focus first on those in the greatest danger.

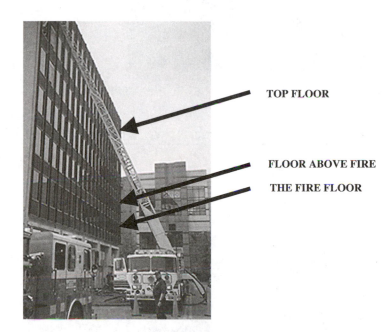

TOP FLOOR

FLOOR ABOVE FIRE

THE FIRE FLOOR

THE MOST LIKELY PLACES TO FIND VICTIMS

Search teams should be assigned to a specific area, and they should use a standard search pattern. When entering a room to begin the search, you must turn either left or right. This turn sets up the basic search pattern. As the search continues, you continue turning in the same direction. This allows you to get back out by reversing your direction, taking you back to where you entered.

When searching in large buildings, consider checking the floor below the fire to learn the floor plan before going to the fire floor. When searching in large or confusing areas, following the walls can take too much time to reach an exit. In these situations, a search rope should be tied off at the entrance where the searchers enter. The rope is played out as they advance, and firefighters can clip onto the rope with their personal safety ropes to search away from the lifeline. This system keeps the firefighters oriented and gives them a quick way out. Thermal imaging cameras can be very helpful finding victims in these low visibility situations. Firefighters must work in pairs when using a thermal imaging camera, because it is very easy to become disoriented while looking into the camera. One team member must keep track of their location and their safety in the building.

WHEN SEARCHING FOLLOW A SEARCH PATTERN

 __Use search ropes when searching in large or confusing areas.__

THERMAL IMAGING HELPS FIND VICTIMS

Primary Search

The primary search is a quick search usually performed before the fire is controlled. Search in beds, under beds, behind sofas and furniture, inside closets, open floor areas, and any place where children could possibly hide.

It is a "do the best you can" search considering fire conditions. When searching, use a simple standardized marking system such as caulk or tape to indicate those areas that have been searched, to avoid duplication of effort. If the search area is large or difficult, notify the

78

Sector Officer or Command when additional search teams are needed. On big search jobs, a Search Sector Officer should be in charge of the operation. This Sector Officer does not search but closely supervises the search operation.

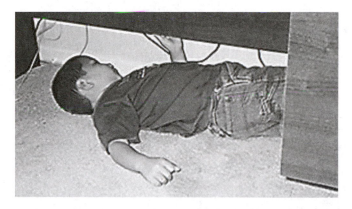

SEARCH WHEREVER CHILDREN COULD HIDE

Search teams should ventilate windows as they come to them, because ventilating improves safety and visibility and increases the chances of finding victims quickly. Venting windows as they go also makes possible exits for the search team if things go wrong. Venting also allows outside crews to track the location and progress of the inside search teams. When victims are found, they are normally dragged out. One night, we had a basement fire in a three-story house. As the attack crews advanced, they found two victims that they dragged out of the basement. They went back in and pulled out two more. A heavy rescue squad went into the basement and came out with three more victims. Before it was over, nine people were dragged out of the basement in five separate searches. At the same time six more people were being rescued from the upper floors. **Don't stop your searches when you find somebody and don't take anything for granted.**

Upon completion of the primary search, results are reported to the Sector Officer or Command, for example: *Search to Command, primary search complete— two victims removed.* When fire and heat conditions make it impossible to complete the primary search, report this to Command.

DON'T STOP YOUR SEARCHES
WHEN YOU FIND A VICTIM
Courtesy of Neil Brake, Tuscaloosa News

Secondary Search

The secondary search normally begins after the fire is extinguished. Take nothing for granted and search one room completely before moving to the next room. The message: *Secondary search complete and negative* guarantees that there is no possibility of a fire victim in the building. It also means that the perimeter around the building was thoroughly checked for victims who may have jumped out a window before the fire departments arrival. I tell my truck officers that the secondary search is for accuracy, not speed. I carefully explain that if they missed anybody, I was going to sue them and that they may wind up bankrupt. They knew that I was kidding, but I noted a marked increase in diligence. Thoroughness is most important in the secondary search. **BE SURE.** There have been cases where the fire investigator finds a body. If Command is not getting search reports, searching is probably not getting done.

 ## *Search reports must be routinely given at every working fire.*

Search Safety

No member should search alone.

There are several principles for safe searches that deserve to be discussed in detail. Before beginning the search, take a few seconds to make an outside size-up of the building. Make a mental note of each of the following:

- Size and shape of the building

- Location where smoke or fire is showing

- Victim locations, if known or suspected

- Possible firefighter escape routes such as fire escapes, ladders, adjacent roofs, porch roofs, and windows

- Obstacles to escape routes such as plywood or security bars over windows, air conditioning units in windows, etc.

Normally if you can't attack the fire, don't search. If there are not enough firefighters on the scene to attack the fire and search at the same time, the search team is in great peril. **Conducting a search when the fire is being attacked is risky, but searching while the fire is allowed to burn borders on suicidal.**

Too many firefighters are killed or injured attempting to rescue people who aren't there. Sometimes the victims were already safe, and sometimes they were never there. When you are told that people are trapped, it is important to find out who you are talking to and how credible they are before making search decisions. We need to ask the right questions such as, "Who are you?" "How do you know that everybody is out?" Or, "How do you know that somebody is trapped?" Don't attempt Hail Mary rescues based on guesses from an unknown bystander. Too often we don't take the time necessary to evaluate reports that someone is trapped. We sometimes claim that we don't have time to ask questions, but we seem to have time to risk our lives in often futile searches.

The risks taken during search must be in proportion to the *realistic* chances of saving a life. Use good judgment. If the homeowner tells you that everybody is out, take no chances during your search. At one fire, the occupants told firefighters that there was no one in the house. Nobody paid any attention to them and Command was not notified. There was a delay in getting water on the fire, resulting in a flashover that almost engulfed the inside search team. A week later, the search crew learned that both homeowners confirmed that there was no one inside the house.

Following a fire that caused firefighter fatalities, the fire chief of an adjacent department had this to say about the tragedy. "The firefighters who died didn't weigh the odds or consider their options. They made the decision to charge in—whatever the consequences, because somebody might need their help." The chief seemed to say that that firefighters shouldn't weigh the odds, and that dying is heroic if there is a potential rescue. No doubt the chief was overcome with grief, but emotion aside, it was the wrong message.

Search teams of two or more should be used for efficiency, speed, and safety. Search teams should be equipped with portable radios, lights, and forcible entry tools. Thermal imaging devices can be very helpful. When searching above the fire, hose lines should be used to protect the search operation.

Some fire folks advocate that firefighters search bedrooms and smaller rooms alone, but **No member should search alone.** While this may work during routine searches, it is not a good idea during difficult searches when heat is high and visibility is impossible. We should prepare for the worst circumstances, not the easy ones. When we split up to search alone routinely, we will most likely split up when the going gets tough.

DO NOT SEARCH ALONE

Entering a building by ladder to search above the fire is very dangerous and is normally used only to search for known victims in specific rooms, not for random searches. In such searches, teams enter with a charged hose line for their protection, close the room door if possible and search the room. It is too dangerous to advance out of the room into a hallway to continue searching the rest of the upstairs. Command must coordinate the search team with the interior sectors for the safety of the search team. When the search of the room is complete, the team retreats back down the ladder the same way they came up.

81

Protect-In-Place When You Can

Sometimes the occupants are safer in their room than in smoke-filled hallways and stairways.

Getting everybody out works well at the typical house or small apartment building fire. For fires in major buildings, the occupants are often safer in their rooms than in smoke filled halls and stairways. A mass evacuation under these circumstances may increase the danger to occupants. At the MGM Grand Hotel fire in Las Vegas, 85 people died. Two-thirds of the victims were found in hallways or on the stairs. Most of those who did die in their rooms, were in rooms located near the elevators shafts, which funneled smoke upwards. It is important to find out just what the situation is, move those victims who are actually in danger, and leave those who are better off in their rooms and provide instruction and reassurance.

Protect-in place is easier said than done because people tend to self-evacuate from fires, and getting the message to them can be very difficult. Announcement and alarm systems can sometimes be used to notify specific floors. It may also be possible to give instructions to people at the windows, using portable bullhorns from the street.

Don't Make Simple Rescues Hard

Protect-in-place and inside rescues are much better than ladder rescues.

One night there was a working fire on an upper floor of an apartment building. There were people at the windows, and many aerial ladder rescues were made. The rescues looked great on the six o'clock news. The problem was that most of the rescues were unnecessary. It turned out that most of the people on the fire floor were better off staying in their apartments, and most who did need to be evacuated could have used stairways. The majority of people "rescued" were placed in more danger than if they had remained in their apartments. Protect-in-place and inside rescues are **much** better than ladder rescues. To make the right rescue decisions, Command must have good information and must coordinate the interior forces with the outside truck companies. **Don't take people down ladders unnecessarily.**

Screwed Up Search And Rescue

One day, I was in charge of Truck 8 at a fire in a garden apartment. There were two women leaning out the third floor window yelling for help. There was little smoke coming from the window, and I could see that the fire on the first floor was being extinguished. I knew the building well, and I knew that the women were in no danger. I told them that they were all right, that they should just stay where they were for a few minutes, and then I continued with my truck work. Soon the fire was out; the women closed their window, and it was over. Later it dawned on me that while I knew that the women were not in trouble, they didn't know it. They could have panicked and jumped. I learned again not to take anything for granted. What I should have done was put up a ladder and gone up to reassure them that they were all right.

I recall a fire that occurred years ago, involving the first floor of a two-story house. There was a solid report of children trapped on the second floor. I was with the engine, and we were assigned to extinguish the fire. After we extinguished the fire, we found two kids in an upstairs front bedroom who were beyond saving. The windows to the room faced the street, and it may have been possible to ladder the windows, advance a search team into the window with a charged line to search the front room. I don't know if it would have made a difference, but it should have been tried.

Recently, a similar fire occurred in another fire department's jurisdiction, with the fire on the first floor and a solid report of a child trapped on the second floor. Neighbors were trying to place a ladder to the child's bedroom window when the fire department arrived.

The fire department stopped the neighbors from their attempt, but never really began a search effort of their own. The engines knocked down the fire on the first floor, and the child was found dead on the second floor right where she was reported to be. Some of the firefighters thought that the child could possibly have been rescued if a search team had entered the room from a ladder with a hose line. These two fires were twenty years apart and occurred in different fire departments, but they read the same. In fires with solid reports of victims in a specific location, search teams should be assigned for rescue.

THE TRUTH ABOUT ROOFS

We have often heard that roof ventilation is vital, and that the best place to open a roof is over the fire. Let's call a time out here and think about what we are doing. Why would we want our firefighters cutting holes in the roof while standing over the fire? We are well aware that many firefighter near misses, injuries and deaths occur during roof operations. The roof of a burning building is obviously a very dangerous place for a firefighter to be. Let's examine our roof ventilation strategies; what are we doing; why we are doing it; and the risk factors.

Australian Roof Venting

I once had an opportunity to talk with a fire officer from Australia. We started comparing how fire departments operate on different continents. Most things were quite similar until we got to roofs. He stated, "We normally don't do roofs." Their philosophy is that it is dangerous to put firefighters on the roofs of burning buildings, and that in most cases the risks outweigh the gains. In this country, we do a lot of roof work, while Australia does very little. Who is wrong? Maybe we both are. We tend to do too much roof work, while they may do too little.

Get Off The Roof

An article in *Firehouse* magazine was entitled "Time to Get Off the Roof," and it showed a series of ten photos of the very narrow escape of three very lucky firefighters from the roof of a dwelling as fire nearly enveloped them. Shortly after the article appeared, I discussed this article with a firefighter. He said: "What really scared me was that while looking at these photos at my fire station, some firefighters who saw the pictures thought it was a normal situation. They didn't realize that it was a near disaster until they read the accompanying article, and that says a lot about our roof work problems."

Avoid Kamikaze Roof Work

Common sense tells us that it is crazy to risk crews by standing over a fire. It's like sawing off the tree branch that you are standing on. Tradition plays a role in our roof ventilation strategies. Times change and we need to adapt. We should not have firefighters with saws and axes slipping and sliding on roofs in heavy smoke, trying to make holes before the fire burns through the supports they are standing on. Slate and tile roofs may appear sound from above, but the rafters below could be burned away, with no indications of a problem. We have been told to look for bubbling tar, melted snow or dry spots on a wet roof, so you know where to open the roof. What these signs should be telling you is to get off the roof, and report the dangerous conditions.

AVOID WORKING OVER THE FIRE

The following statements were made by firefighters about actual fire operations. They probably sound familiar because these mistakes are common at major fires. If you re-read them slowly and ask yourself what was happening and why, you will realize that firefighters' lives were risked trying to save property that was already wasted.

"Conditions worsened, the cockloft was now fully involved and Command knew that the only chance to save the building would be to open the roof immediately."

"I started cutting and I looked back and saw the roof sag. Within moments, the fire vented itself and several firefighters had narrow escapes off the collapsing roof."

"After roof ventilation was ordered, we went up the aerial ladder to the roof and we were just stepping onto the roof when the rear half of the roof collapsed. The roof was supported by steel bar joists."

These statements show that we sometimes have the bad habit of working over trusses involved in fire. Beware of wide span roofs because they are usually supported by steel or wooden trusses that quickly fail under fire conditions. **If there is a working fire involving a wide span roof, expect collapse, stay off the roof, and get everyone out from under it.**

🔑 **_Do not stand over fires and cut the roof that you're standing on._**

House Roofs: Fire In The Attic
If the fire is in the attic, it is too dangerous to be playing on the roof.

Some departments believe that you only need to cut roof ventilation holes when there is fire in the attic. **Time out!** Think about this for a minute. What this means is if there is no fire in

84

the attic, stay off the roof, but if there is fire in the attic, stand over the fire, and start cutting the roof. Cutting the roof directly over the fire is very risky, and the rewards are minimal. Residential roof sheathing is flimsy at best. A fire in the attic or cockloft will burn through and vent itself without our help. Roof cutting from roofing ladders reduces the risks somewhat, but firefighters working over the fire are still vulnerable to falls, roof collapse, and flame erupting through the roof.

House Roofs: No Fire In The Attic

If the fire is not in the attic, it isn't smart to cut open the roof, because the ceilings keep the majority of smoke out of the attic. It is difficult to impossible to push down the interior ceilings from the roof to let the smoke out. Removing hatches and skylights is a very effective way to ventilate these fires. Like all roof operations, check that there is no fire immediately under the roof on which you are standing. Taking the domes off attic fans and removing the tops of attic ventilators (those whirly bird things that spin from natural attic heat) can also help. One night a roof team was venting at a house fire when the attic fan kicked on due to the heat from the fire. The fire blew out of the fan dome 12 feet into the air, scaring the hell out of them. They were unaware that the fire was in the attic.

The bottom line: Don't cut open house roofs. Roof cutting over an attic fire is foolish and dangerous. Cutting the roof when the fire is not in the attic is a waste of time. The best way to ventilate house fires is by using horizontal ventilation and by removing hatches and skylights. Information on fighting attic fires is found in the Fireground Operations chapter.

Smart Non Residential Roof Ventilation

Roof ventilation is often very valuable in these buildings, so a roof team should be quickly sent to the roof. Do easy venting first (not working over an attic fire) by removing the obvious roof openings such as skylights and hatches, and by opening roof stairway bulkhead doors.

Venting through a skylight is a very effective way to ventilate a building. Skylights should be lifted out of the way if possible, and then laid upside down as a warning of the roof opening. If it becomes necessary to break the glass out of a skylight, give warning to avoid injuring those below. After removing a skylight (or hatch), check for a panel or glass below the skylight and make sure it is open.

REMOVING A ROOF HATCH

REMOVING A SKYLIGHT FOR VENTILATION

Proper Cutting Of Vent Holes In Roofs

The key to cutting vent holes in roofs is quick, simple operations.

Ventilation can be accomplished by making a large hole in the roof provided:

- The roof can't be vented the easy way by opening skylights, hatches, or roof stairway doors

- Window ventilation is underway or completed

- The roof crew doesn't work over the fire

Chief Dinosaur says, "If you cut a hole that's not over the fire, it can draw the fire through the attic, you could lose the whole damn attic." The chief is right, you may lose the attic. Of course, if you choose to vent over the fire you could lose your roof crew — not a very good tradeoff.

One person on the roof should supervise the operation for the safety of the roof team cutters. The cutters can be so involved in venting that they are oblivious to changing conditions on the fireground. The key to cutting vent holes in roofs is to keep operations quick and simple. Make one good vent hole and get off the roof. The more people on the roof and the longer they are there, the greater the danger. Towers and buckets can improve safety because truck crews can remain in the bucket while opening roofs. SCBA should be used in the bucket.

Roof Teams

Safety is a primary concern for the roof team. If there is any doubt about the stability of a roof, stay off it. Nobody should be allowed on the roof who doesn't understand building construction. This team should have at least two radio-equipped members for safety and efficiency. There should be at least two escape routes off the roof, and everyone working on the roof should have SCBA. Limit personnel allowed on the roof to the minimum needed to do the job. When roof operations are completed, the roof team should descend by ladder, not the interior stairs.

HAVE AT LEAST TWO WAYS OFF THE ROOF

The roof team is often in an excellent position to see overall conditions on the fireground and should promptly report the roof status to Command as follows:

- Fire, smoke, and stability conditions on the roof

- If roof ventilation is needed

- If they are unable to vent the roof

- When hatches or skylights are opened as well as reporting any smoke or fire issuing from these openings

- When roof cutting is needed

- Fire location as seen from the roof

- Exposure conditions seen from the roof

- Heavy objects, i.e., parapets or roof top air conditioning units that may be a collapse hazard

- The presence of overhead power lines and utilities

- When roof work is complete

Example of a roof team report:

Truck 6 Roof Team to Command: The hatch is off — heavy smoke venting. The fire is in quadrant C (right side, rear) *and exposures B* (left side) *and D* (right side) *look okay.*

The Roof SCBA Dilemma
If you need a mask, you probably shouldn't be on the roof.

The rule is that everyone working on the roof should be using SCBA. The reality is that most roof teams don't use it. One reason is that walking around on a roof using SCBA may be more dangerous than not using it. Visibility on the roof is often poor even without the limits of a mask. Footing is dangerous; there are open shafts, roof edges, and open hatches and skylights. If you need a mask on the roof because of smoke conditions, you probably shouldn't be up there.

If you are doing kamikaze roof work, working in heavy smoke or cutting the roof over the fire, using SCBA is vital. If you are working smart and safe, not working over the fire or in heavy smoke, there is less need for using SCBA. In this case, it may be best to have the SCBA on your back and use the facepiece when you are actually opening the roof or if conditions change.

The Truth About Trench Cuts

Trenching is the opening up of a roof by cutting a trench all the way across the roof to limit the spread of fire. A trench cut can be just what you need in certain situations, if you know when, where and how to make one, and you have the personnel on hand to do it. If you have a fire in the cockloft of a long narrow strip shopping center, the trench cut may save the day and cut off the fire.

The reality is that for most departments the trench cut is a "Hail Mary" play that is seldom used or is unsuccessful. Even in large departments with more resources, it is not used very much and it is not unusual for the fire to pass the cut before it is completed. Everybody talks trench cut, but few actually do it. There are a very few times that the conditions are right to use it, and even then it doesn't always work as advertised.

Recently there was an article in a national fire magazine about trench cutting. The article was illustrated with a photo that showed two firefighters on the roof of a dwelling. They had made a small hole in the roof with a power saw. The photo was captioned "This is a trench cut being made." There were two errors; we don't trench cut dwelling roofs, and the opening they made was a ventilation hole, not a trench cut.

Why Trench Cuts Are Seldom Practical

A successful trench cut requires the following conditions.

- The structure has a long narrow roof sometimes found on garden apartments, strip shopping centers and the narrow throat area of large H shaped apartment buildings. A narrow roof is important for speed of cutting. Wide roofs usually take too long to make the extensive cuts needed

- The fire must be located in the cockloft of the building

- The fire should involve no more than about one-fourth of the cockloft to allow time to get ahead of the fire

- Large numbers of personnel must be available, since trench cuts usually require at least an engine company and two truck companies on the roof, and several engines and a truck in the interior under where the cut is being made. This is in addition to all the other resource requirements needed at a major fire

- Trench cuts are complicated. A high level of expertise is needed to recognize the proper conditions for a cut, to select the right place and method to cut, and to coordinate the operation while providing roof safety. It is very difficult for most departments to acquire and maintain trench cut skills that are infrequently used

Avoiding The Trench Cut

Trenching the top floor ceiling instead of the roof to control a cockloft fire can be very effective. It can be faster, safer, and much easier than roof trenching. To perform the inside trench cut, the fire should involve no more than about one-fourth of the cockloft to allow time to get ahead of the fire. The ceilings are pulled well ahead of the fire from the underside, creating a trench in the top floor ceiling from outside wall to outside wall. Start with a trench several feet wide. Then enlarge it as time allows. The fire can sometimes be stopped by hose lines directed into the attic from the floor below.

SUMMARY

Assign truck crews to specific jobs.

Truck company teams should enter the building together, work together, and leave together as a team, so that no member operates alone.

Vent early and vent often. When the attack crews are ready to go, the vent crews start venting. Quickly break out the windows using ladders and tools. When in doubt break it out, and quickly clear away drapes, screens, and other obstructions to speed the venting. Glass is cheap and fire fighting inside poorly ventilated structures is dangerous.

The fastest and safest way to ventilate windows is to use ground ladders to break out the windows. The aerial ladder can also be very effective to punch out windows to ventilate upper floors.

Proper placement of ground ladders is vital for fireground rescue, for ventilation and for firefighter emergency exits. When firefighters enter a building by ladder, don't remove the ladder until all members have exited.

PPV can work very effectively in many situations, including interior attack, if it is used properly.

Search teams should be designated for search and rescue.

Normally, if you can't attack the fire, don't search.

Search the fire floor first because most victims are found there. The next most likely places to find victims are the floor above the fire and the top floor.

Protect-in-place and inside rescues are much safer than ladder rescues.

Beware of wide span roofs on commercial buildings. If there is a working fire below a wide span roof, stay off the roof, and get people out from under it.

The easiest way to ventilate a roof, is to remove hatches and skylights.

Roof cutting directly over an attic fire is foolish. When the fire is not in the attic, cutting the roof is a waste of time. In both cases use horizontal ventilation and remove hatches and skylights.

When fires are in commercial buildings, ventilate by removing skylights, hatches and by opening roof stairway doors. If more ventilation is needed, cut one large vent hole using quick simple operations.

Safety is a primary concern for the roof team. This team should have at least two members with radios for safety and efficiency and there should be at least two escape routes off the roof. Towers and buckets improve ventilation safety because truck crews can remain in the bucket to open roofs.

8

FIREGROUND OPERATIONS

"Lead me, follow me, or get out of my way." - George S. Patton

This chapter explains fire fighting techniques for fires in basements, attics, partitions, high-rise buildings, and overhaul operations.

BASEMENT FIRES

Life Safety and Basement Fires
Going down the stairs at a working basement fire should be the last option, not the first.

It is still a widely-accepted practice to attack basement fires by descending the basement stairs. You go down and kill it. We are told to go down the stairs backwards, so it will be easier to scurry back out if you start to burn up. **Time out!** The stairways are chimneys for all the heat and smoke. To extinguish a fire in a fireplace, we put out the fire at the fireplace. No one except Santa Claus would consider going down the chimney. Going down the stairs at a working basement fire is like going down the chimney, and should be the last option, not the first.

A fire officer feels his way down the basement stairs through heavy smoke and sees the glow of flames under the stairwell. "Oh God" the lieutenant thinks "Are we going to be able to get out of here?" Much later the lieutenant discovers that the basement has a door leading to the back yard that could have been used for an easier and safer fire attack.

If flashover occurs while advancing down the stairs, or if the gas meters ignite, the crews are in grave danger. Once in the basement, if things turn sour, it may be difficult to find the stairs. Once on the stairs, you have to exit upward right through the chimney. When the exiting crew reaches the first floor, they still would not be safe, because they would be inside a building that is heavily charged with smoke. The bottom line is that going down the stairs is very dangerous.

91

When attacking down the basement stairs, the open basement door allows all the heat and smoke from the basement to fill the structure. This deadly and blinding smoke increases the risks to the search crews and the victims on the floors above, reducing the odds for a successful rescue. The reason these risks are being taken is to reduce property damage in the basement. **It's time to re-think our basement strategies.**

AVOID GOING DOWN THE CHIMNEY

Basement Fires The Smart Way

Typically a basement has an outside entrance. A basic principle is to attack basement fires from the same level, not from above. When there is an outside basement entrance, close the basement door on the first floor if possible to protect both the occupants and the firefighters in the building. Outside crews then attack the fire through the outside basement door with both an attack and backup line. If these attack crews have a problem and need to back out, they can back straight out into fresh air.

Before Command gives the outside crews a green light to advance, interior crews must be in a safe location and aware of the pending attack through the outside entrance. This operation requires strong command and control, SOPs and good communications. The inside engine crew shouldn't sit on top of the fire any longer than necessary.

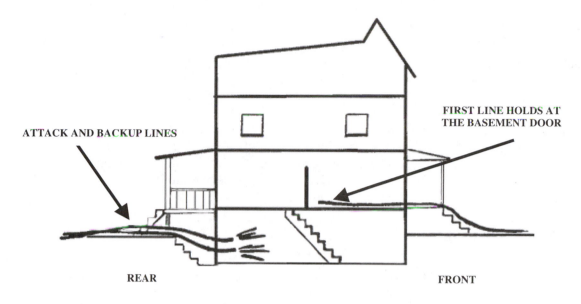

ATTACK AND BACKUP LINES

FIRST LINE HOLDS AT
THE BASEMENT DOOR

REAR

FRONT

BASEMENT FIRES THE SMART WAY

***Make every effort to attack basement fires from the
same level, not from above.***

Basement Fires, Plan B

If there is no outside basement door, an outside basement window can be used to knockdown the fire or we can create a window-like opening. Chief Dinosaur says "Never attack through a window; you always go inside." We usually use interior attack, but not always. Safety is a major concern, and it is certainly safer to achieve knockdown through a basement door or window, then to send a crew down the chimney.

Another safe way of extinguishing a stubborn basement fire is to drown it. Master steams into the first floor have a sprinkler-like effect on the fire. It's crude, but if you fill the basement with water, the fire disappears. This can avoid "Hail Mary" plays like making repeated futile efforts to go down the basement stairs or trying to cut holes over the fire.

When Going Down The Basement Stairs Is Okay

Going down the stairs is okay when it is reasonably safe, that is, when there is little heat or smoke and the fire is obviously minor. A good rule of thumb is: when you can see the basement floor and there is little or no heat, it is safe to go down. It is also safe when the stairway is enclosed and the basement level door is closed and not hot.

93

Basement Ventilation

Don't stand over the fire cutting holes in the floor.

Good ventilation is essential in any basement or below ground fire. Quickly vent all basement windows, including those at ground level and in window wells. Any basement level doors to the outside should be opened up.

Venting a basement fire by standing over the fire and cutting holes in the floor is dumb and a good way to drop into the fire. The occupants are usually out of the building or already lost long before we even think about cutting holes in the first floor. There is almost never any life to be saved in a heavily involved basement fire. The only reason that we can justify risking our own lives is when there is a reasonable chance of saving a life.

Quickly vent basement fires.

AVOID WORKING OVER A BASEMENT FIRE

The Floors Above The Basement Fire

When faced with a serious basement fire, plan for the possibility that the fire could get away from you. The basement has ducts, wire chases, and partitions going upward throughout the building. Fire often spreads up through these routes, so it should be SOP to have companies above the fire quickly open baseboards, walls, and ceilings to cut off the fire. It is important to keep check on the attic or cockloft, since the basement fire can easily spread there.

Basement Fire SOPs

Basement fires are dangerous and should be handled using SOP guidelines to prevent unnecessary injuries. Communications and coordination are very important in the proper handling of basement fires. A sample basement fire SOP follows.

The First Engine

The first engine company lays and charges a hose line into the first floor of the fire building at the interior basement stairway door. The purpose of this hose line is to protect the search and rescue operation, and to confine the fire to the basement. This company does **not** attack the fire using the interior stairs, unless the fire is obviously minor. They should close the door to confine the fire. If they are unable to close the door, and the fire is coming out of the door, they can put the nozzle on fog and aim it at the stairway to prevent the fire from spreading. They should use care not to direct a line into the basement or attempt to extinguish the basement fire. A backup hose line should protect the crew on the first floor.

If it is unsafe for the engine company to hold a position on the first floor, they should notify Command to protect companies that may be operating above the fire.

If search and rescue is completed before the basement fire is controlled, the first engine company must get off the first floor for their own safety. They should back out of the building and remain ready to advance from a safe position. No one should stay on the floor above a basement fire any longer than necessary.

The Second Engine

The second engine company attempts to locate an outside basement entrance. Upon finding an entrance or suitable window, they should prepare to attack the fire. They do not attack the fire without clearing it through Command. When attacking through an outside basement entrance, they should consider using a straight stream. This will give greater penetration into the fire area, and is safer for the attack crews.

The Third Hose Line

The third hose line backs up the engine company that is attacking the fire through the outside basement entrance. (typically the second engine)

The Fourth Hose Line

The fourth hose line backs up the engine company that is protecting both the first floor and the search operation. (typically the first engine)

The Truck Company

All basement windows should be taken out, and outside basement doors opened to allow the fire and heat to be vented to the outside. If there are no doors and windows, the truck company should create them. The truck also ventilates the upper floors, performs search and rescue and opens up the partitions.

Basement Fire Summary

A basic principal is to attack basement fires from the same level, not from above. Don't go down the stairs at working basement fires if you can avoid it, because the stairways are chimneys for all the heat and smoke.

Most basements have an outside entrance or a basement window that allow an outside crew to safely attack the fire after getting approval from Command.

Good ventilation is essential at basement fires. Quickly vent all basement windows, either at ground level or in window wells. All basement level doors should be opened up.

ATTIC FIRES

"Hand me the pipe." the captain yelled down through the attic opening. His crew handed him the playpipe, and that was the last time they saw him alive. The attic was not fully vented, the heat built up, and he couldn't find his way out.

We should avoid sending hose line crews into attics because the risks are high and the gain is property. The hazards are flashover, entanglement, unsure footing, and disorientation. There also is usually only one exit that may be hard to find. There are usually better and safer ways to extinguish attic fires as the following methods show.

An excellent attack method is to pull the top floor ceiling in selected places, and insert revolving distributors into the attic from below. To do this, place the revolving distributors on playpipes by using short extension pipes. The results will be similar to large sprinkler heads operating in the attic. This method can be enhanced by an **un-vented** attic. If you think about it, this may be the best option, because it is safe and does not require roof work or major ceiling pulling. Some departments use piercing nozzles, which are inserted into the attic through the top floor ceilings. A similar, but less effective method is to pull top floor ceilings and direct attack lines into the attic from the top floor. These indirect attack methods require changes in our attitudes and possibly our equipment.

**EXTINGUISHING AN ATTIC FIRE USING
A REVOLVING DISTRIBUTOR**

Another good way of extinguishing attic fires is to direct a hose line or elevated stream through an attic window or dormer. If there is no such opening pull out the attic vents or cut a hole in the side of the attic (gable ends), and direct hose streams into the attic. This is safer than risking firefighters inside the attic or over the fire. It is important to know the types of attic construction in your area to know which fire fighting methods work best.

Indirect attacks are safer and smarter than sending crews into the attic.

Plan B, Attic Interior Attack

If the attic fire is not vented, stay out of the attic

Interior attack can be useful in some situations, such as when the fire is minor or when other methods won't work and the fire is well vented. Before entering the attic, you must first check the top floor thoroughly because many attic fires originate on the top floor and extend up into the attic. You don't want to be in the attic with fire below you. To begin an interior attack, the engine company looks for an attic entrance, usually a stairway, pull down stairs or a hatch. Try to extinguish the fire from where entry is made into the attic, so you are close to an exit. When attacking in an attic, use a backup line, place a light at the attic exit as a guide to mark the way out and have a RIT standing by near the attic entrance for firefighter rescue.

The key to successful extinguishment is attic ventilation. If the attic fire has not vented, stay out of the attic because it's too dangerous to advance attack lines in unvented attics. At house fires, ventilation can usually be safely accomplished from the gable sides of the attic by pulling out the vents or cutting a ventilation hole in the side of the attic.

Avoid Fighting Attic Fires From The Roof

One afternoon we had a fire in the cockloft of a building. There was no access to the cockloft and the buildings construction would not let our hooks open the top floor ceilings. Since nothing else seemed to work, we went to the roof to see if we could get to the fire from there. It was a difficult and dangerous operation that didn't work very well. Fighting the fire from above meant that we were working over the fire; this was not smart. Think about it, you are risking the lives of those over the fire for what? To save part of a burned-up roof? It really doesn't make sense.

Attic Fire Summary

An excellent attack method for attic fires is to insert attack lines equipped with short extension pipes and revolving distributors into the attic. No attic ventilation is needed. The result is similar to large sprinkler heads operating in the attic.

Another method of extinguishing attic fires is by directing a hose line or elevated stream through an attic window or dormer.

At house fires, ventilation can usually be safely accomplished from the gable sides of the attic by pulling out the vents or cutting a ventilation hole in the side of the attic.

Avoid sending hose line crews into attics because the risks are high and the gain is property. If attack lines must be used in the attic, the attic must be well ventilated. If a working attic fire has not vented, stay out of the attic.

When attack lines are used in the attic, a backup line must be used, the way out must be marked and a RIT crew must be at the entrance for the safety and orientation of the attack crews.

PARTITION FIRES

The key to success is aggressive truck work.

Fires in the partitions, walls, ceilings, and duct chases can be very difficult to control. The key to success is aggressive truck work. This type of fire needs a lot of firefighters with hooks to open up the concealed spaces. It is important that the truckies, or those assigned to truck work, do truck work and only truck work. Opening up quickly with hooks while the engines knockdown the exposed fire works best. The direction of the fire's spread is somewhat predictable. If the fire is in a wall, it is likely to run up the wall towards the attic and may sometimes go across the ceiling. Get ahead of the fire. Don't concentrate on where the fire is, focus on where it is going. When the fire is in a duct, (normally found in restaurants or in heating and cooling), first shut off the system. Position hose lines at both ends of the duct and check out the entire length of the ductwork.

In the middle of a serious fire in the partitions of a house, another officer wanted my truck company to crawl around in the attic looking for fire extension. I said, "No way," because we would be in the attic without a line and without knowing where the fire was below us. Safety is scary at these fires because the fire location is often not well known or defined. The fire seems to be here, there and everywhere, and sometimes it is. All officers should be super careful not to get into situations where they could be cut off and trapped by the fast spreading fire. Open up partitions quickly and closely monitor fire conditions.

TRUCKS OPEN UP AND ENGINES EXTINGUISH PARTITION FIRES

HIGH-RISE OPERATIONS

High-rise buildings used to be confined to the downtown section of big cities, but no more. Many fire departments now have high-rise buildings in their towns.

There are entire books written on high-rise fire fighting operations. What follows is a very general guideline to point you in the right direction and cover some high-rise fire fighting basics.

High-Rise ICS

The name of the game is early fire control.

High-rise ICS should be started as early as possible, but not if it reduces the effectiveness of the operation. Use your first alarm assignment to confine and extinguish the fire and be sure to supply sprinklers and standpipes. Some departments assign the first due engine to lobby control. Although lobby control is an important function, the name of the game is early fire control. The idea of setting up the high-rise ICS structure early in the incident is good, but not at the sacrifice of rapid suppression. If the first companies puts out the fire, you may not need lobby control. If they don't, you may need lobby control for the next three days.

Some fire departments send heavier responses to high-rise buildings so that the first alarm response has enough resources for both suppression and the start-up of high-rise ICS. The Philadelphia Fire Department sends an additional company above the normal assignment. The Montgomery County, Maryland Fire Department dispatches five engine companies, three truck companies, a heavy rescue and two chief officers to reported fires in high-rise buildings.

High-Rise Fire Fighting

Upon arrival, elevators should be placed under manual control and returned to the lobby for fire department use. The first engine checks the fire floor and promptly reports conditions to Command. This engine company locates the fire, connects to the nearest standpipe on the floor below and attacks the fire. It is usually a mistake to connect to the standpipe before finding the fire, because smoke and distances can be misleading, and precious time may be lost. It is important that companies be familiar with building layouts and their sprinkler and standpipe systems.

The first line is very important, and often requires more than one company to put it into action. Consider using 2½ -inch lines for fire attack if you have the resources to do it. The second line backs up the first line, ensuring more firepower and improving safety. If the initial attack hose lines are not advancing, scout around for another way to attack the fire. An attack from another direction may be more effective, but must be closely coordinated with the initial attack lines to avoid opposing hose lines.

The third line protects exposures on the floor above the fire. The fourth line either assists with fire attack, or protects exposures on the fire floor, depending upon the circumstances. Keeping a single hose line operating requires three or four engine companies or equivalent personnel: one or two engines operate on the hose line, one is ready to relieve and one is resting up.

The building's ventilation system is normally shut down until the location of the fire and the smoke conditions in the building are known. In many new buildings, the ventilation systems will automatically shut down when the fire alarm system is activated. Removing windows for ventilation is difficult and dangerous, and should only be used as a last resort. Truck companies perform search and rescue. The floors above the fire are usually grouped together. For example, floors ten through twenty may be placed in the Ten/Twenty Sector, under the command of a chief officer.

Usually it is best to leave high-rise occupants in place if possible. The corridors and stairwells are often smoke filled, while rooms are usually relatively smoke free. When evacuation is necessary, determine which stairways will be used for evacuation, and which will be used for advancing fire crews. Sometimes, people who are self evacuating upon our arrival will determine stairway usage.

High-rise fires require a whole lot of resources for suppression, rescue and logistics. Many fire departments are not fully aware of the personnel requirements, and even big departments tend to underplay the high-rise fire. Most fire departments do not have the massive resources needed for a major high-rise fire, and must rely on mutual aid. **The moral in high-rise fires: Call for lots of help fast.**

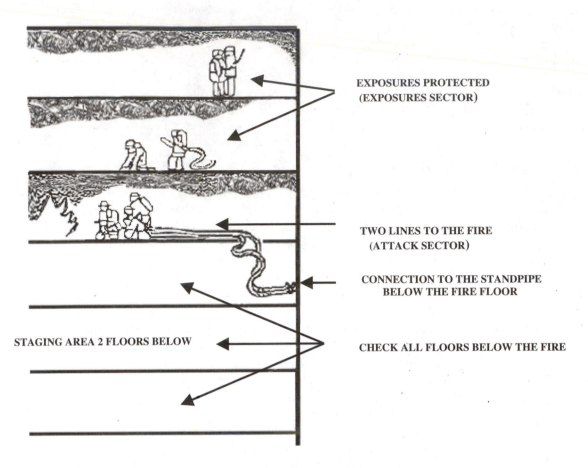

EXPOSURES PROTECTED
(EXPOSURES SECTOR)

TWO LINES TO THE FIRE
(ATTACK SECTOR)

CONNECTION TO THE STANDPIPE
BELOW THE FIRE FLOOR

STAGING AREA 2 FLOORS BELOW

CHECK ALL FLOORS BELOW THE FIRE

HIGH-RISE OPERATIONS

High-Rise Command

One night there was a fire in a high-rise college dormitory. The incident commander went into the building. Willy, the local fire buff, stood across the street and watched as things began to go downhill fast. Soon, Willy saw students begin to appear at the windows. The smoke worsened and people began to shout and lower sheets tied together. A tragedy was clearly in the making and suddenly poor Willy seemed to be the only one aware of it. The IC and most of the firefighters were inside. The outside firefighters were busy charging standpipes and the like. When Willy tried to tell a few firefighters what he saw, they answered, "Shut up and get out of the way, Willy." Seeing a desperate situation, Willy made a desperate decision. He went to an apparatus radio and, using fire department terminology, called for a second alarm, pretending that he was the chief. They needed the help that Willy had called for, but he broke a big rule and poor Willy had to go into hiding for a while. The moral of this story is that the incident commander must have a view of the incident.

Establish the command post out of the collapse zone and an operational command post is then established several floors below the fire to coordinate the operation. Pre-plans and floor plans should be used, and the building engineer (who knows more about the building than anyone else) should be at the command post. If the engineer and maintenance crews are not present, call for them.

100

Establish a staging area at least two floors below the fire, so that resources will be quickly available. A second staging area is set up several blocks from the incident where incoming companies initially report.

Chief officers are usually assigned to supervise sectors. Establish a Safety Sector because a single Safety Officer cannot effectively handle a serious high-rise fire. Accountability, collapse, rehab, and falling glass are only a few of the safety problems that make close supervision necessary. Search lines may be needed because high-rise buildings often have maze-type cubicles, long hallways, and large open spaces—your basic firefighter's nightmare.

Communications

Communications are often a problem in high-rise buildings. It is often best to divide the communications between several radio channels. For example, channels can be established for fire attack, exposures, command and logistics. A communications officer must be used to coordinate the radio operation. Use telephones and cell phones to reduce radio communications, particularly between Operations, Command, and Staging. When radio problems exist, radio relays or bullhorns can be used in the stairwells to communicate with the firefighters.

CELL PHONES REDUCE RADIO TRAFFIC

Elevator Operations

Most elevators prominently display signs stating: DO NOT USE ELEVATORS IN THE EVENT OF A FIRE OR EMERGENCY, USE THE STAIRWAYS. This may be telling us something. Taking an elevator in a building that is on fire is risky. We have had too many firefighter near misses, injuries, and deaths from careless elevator operations. Rule 1: don't use an elevator if you don't have to. If it is practical to walk up, do it. Consider it a part of your physical fitness program. An exception is Empire State Building type fires, when the height of the fire precludes using the stairs. When you must walk up remember that a fully equipped firefighter can climb at the rate of about a floor a minute. Crews must pace themselves so they are capable of fire fighting when they get to the fire floor. Command must factor the climb time into incident strategy.

Dumb Elevator Operations

One night an alarm came in, reporting smoke on the fourth floor of a museum. The museum had a history of malfunctioning fire alarms. There had never been a working fire in the building before, and there was little reason to think that there would be one this time. Nothing was showing, and the guards were not excited. There were two banks of elevators, one to the left and the other to the right. The first company randomly chose the elevators on the right. They went into the elevator and pushed the button for the fire floor. When the doors opened, they were met by a wall of choking smoke and high heat. Quickly they pushed the down button. Fortunately, the door closed and they were able to get back down. They were doubly fortunate because the other bank of elevators opened directly into the fire.

At another fire, an engine company responded on the second alarm for a serious fire in a commercial building. The fire was on the third floor. They entered the elevator and incredibly pressed three, the fire floor. As the elevator rose it began to fill with smoke, and a police officer who was on the elevator said 'I'm out of here," pressed floor two and got off. The engine th en pressed three again to take them to the fire floor. When the door opened, they were immediately in big trouble. They were unable to get the elevator to go back down, so they were forced to enter the fire floor searching for a way out. They were rescued from windows by portable ladders. Why do these dumb things happen? A big reason is that we become accustomed to using elevators for alarm bells, smells, and routine business, so when there is a fire, we tend to get on the elevator as we usually have done.

Safe Elevator Operations

Use elevators that are equipped with a firefighter's control feature. Too often firefighters fail to place elevators into the fire service control mode. This happens because firefighters are not used to this mode of operation. When elevators are used, companies should not use elevators until the first due company reports the elevators safe to use.

The first company to use an elevator should adhere to the following guidelines.

- Notify Command that you are using an elevator

- Note the relationship of the elevator to the stairway

- Shine a strong light up the shaft between the outer and inner doors to determine if smoke is entering the shaft at some point above. If there is smoke visible in the shaft, don't use the elevator

- Wear SCBA with cylinder valves open and PASS devices activated

- Carry forcible entry tools; they can be used to pry open the elevator doors if the elevator fails to stop or controls don't work

- Never overcrowd the elevator

- When going up in an elevator, stop the car periodically to:

    — Test the elevator controls

    — Shine a light up the shaft to check for smoke

- Stop the elevator at least two floors below the fire floor

- When getting off an elevator, don't let the elevator go until you are sure that you will no longer need it

- Notify Command when elevator use is safe

- Promptly report the conditions on the fire floor to Command

Good elevator procedures should be a part of your fire departments SOPs, and then practiced routinely for your safety and survival.

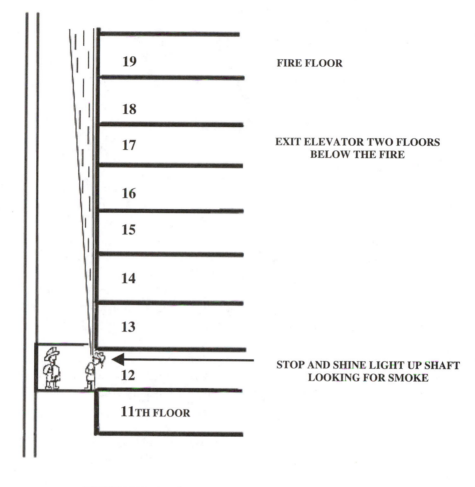

19   **FIRE FLOOR**

18

17   **EXIT ELEVATOR TWO FLOORS**
                   BELOW THE FIRE

16

15

14

13

12   **STOP AND SHINE LIGHT UP SHAFT**
                   LOOKING FOR SMOKE

11TH **FLOOR**

ELEVATOR OPERATIONS

High-Rise Logistics

Because of the time and distance factors in fighting high-rise fires, it is important that companies take in all the equipment they may need. "Try to take the truck with you when you go in," is the advice of one veteran truck officer. "If yo u don't have something that you need, it can take forever to get it." First arriving companies should bring a variety of equipment. Later arriving companies should be notified of the logistical needs so they can take the proper equipment up with them.

Good logistics can often make the difference between the success and failure of an operation. Air supply is a key element of logistics. Consider raising an air fill line up the side of the building to fill the bottles on site. Sometimes an aerial tower can be used as an elevator to shuttle equipment to an upper floor. If logistics must be supplied by using the stairways, the rule of thumb is: one logistics person for each two floors. Logistics should be located near the operations command post, several floors below the fire.

AIR SUPPLY IS OFTEN THE KEY TO HIGH RISE LOGISTICS

If The Hand Line Attack Fails

If the fire is on a lower floor and cannot be extinguished by hand lines, it may be possible to use elevated streams to knockdown the fire. The Washington, D.C. Fire Department, among others, has used this method successfully. One example was a fire on the fourth floor of an office building. An interior attack had failed, and fire was coming out of about a dozen windows. The fire was rapidly knocked down by a ladder pipe stream. At another incident, the fire was on the seventh floor, which is normally above the effective reach of elevated streams. Fortunately there was an elevated highway ramp next to the fire building. An aerial tower was positioned up on the ramp and was used to extinguish the fire. If the ramp had not been there, the fire would probably still be burning.

Before beginning the exterior attack, account for all personnel. Interior companies can retreat to a safe position within the fire building (usually two floors below the fire.) This fallback position is not used at fires in dwellings or buildings of ordinary construction; it is only used for fires in high-rise buildings.

When the fire is above the reach of elevated streams, consider placing monitor nozzles in service on the fire floor if the water supply is adequate. This may give the extra water needed to control the fire.

Giving Up

When the fire is on more than one floor, rescue is complete and your best efforts are not working, it is time to consider giving up. I know, fire departments are never supposed to quit, but we must look at the risk against the gain. The magnitude of these fires can be beyond the capability of any fire department to extinguish. We can't lose lives in a futile effort to save property. Giving up on the high-rise fire has been done, and it is the only reasonable option in some cases.

High-Rise Summary

High-rise ICS should be started early, but not at the sacrifice of rapid suppression.

The first hose line is very important, and often requires more than one company to put it into action. During ongoing operations each hose line will require three or four engine companies (or equivalent personnel.)

High-rise fires require tremendous numbers of personnel. Call for help quickly and in force.

It is usually best to shut down the building's ventilation system and avoid venting until the location of the fire and the smoke conditions in the building are known.

Try to leave high-rise occupants in place because the corridors and stairwells are often smoke filled, while rooms are usually relatively smoke free. When evacuation is necessary, determine which stairways will be used for evacuation, and which will be used for advancing fire crews.

The main command post must be located with a view of the building. An operational command post is then established several floors below the fire to coordinate the operation.

Establish a staging area at least two floors below the fire, so that resources are quickly available.

Command should establish a Safety Sector at serious high-rise fires.

Elevators are dangerous places to be during fires. Don't use them if you can avoid it. If it is practical to walk up, do it. When companies must enter an elevator, they should use elevators equipped with the firefighter's control feature. Other companies should not use elevators until the first company reports that the elevator is safe to use.

Proper logistics can make the difference between the success or failure of an operation, and effective air supply is often the key to success.

If the high-rise fire cannot be extinguished by hand lines, and the fire is on a lower floor, it may be possible to use elevated streams to knockdown the fire.

When the fire is on more than one floor, rescue is complete and your best efforts are not working, it is time to consider giving up. We can't afford to lose lives in a futile effort to save property.

OVERHAUL

There are many building foundations that stand as a tribute to the failure to open up.

Too often an engine company does a good job of fire control, and after knockdown, will stand around congratulating themselves on what a good job that they did. Meanwhile the fire may be burning undetected in the ceiling or walls or have worked its way up into the attic. It's easy to believe that everything must be okay if you can't see fire anymore. **WRONG!**

It is critical to make sure that the fire has not spread. A knowledge of building construction techniques is important to understanding fire travel during overhaul operations. Thermal imaging cameras should assist with overhauling if available. Concealed spaces should be opened up if there is suspicion that fire has entered them. Check all walls above the fire, including the attic. Trucks should bring in extra hooks because sometimes the engine crews will be able to help them open up.

Some believe that pulling ceilings and opening up walls, pipe chases and partitions cause unnecessary damage. The truth is the opposite; not opening up can cause major losses. Open it up or lose it. Working fires do a great deal of damage and almost everything will be ripped out by the repair contractors, so don't be afraid to open up to ensure that no fire remains. No owner ever lost a building from the aggressive opening up of partitions, but there are many building foundations that stand as a tribute to the failure to open up.

Command should organize overhaul companies by floors, or areas of floors depending upon the size of the building. Overhaul is often a bigger job than it appears, especially when the troops are exhausted. Remember that many hands make light work. Consider the weather, the condition of the troops, the condition of the building, and the needs of fire investigation.

Don't forget to consider the safety of fire investigators, because they sometimes lack proper safety equipment or take unnecessary chances.

Be sure there is no fire remaining, particularly in structures that will be reoccupied after the fire department leaves. The IC need not look at everything personally. When there are places that are difficult to check, the IC should pick a solid and reliable officer to check it out, then follow up by having other officers check it periodically. Different people see different things. Time is on the side of the fire department during overhaul operations. Take advantage of it.

On the other hand, do not make overhaul a bigger job than necessary, particularly when it comes to the burned furnishings in the building. One large fire department used to remove everything and wet it down outside. When their fire activity increased, they found that they didn't have the time or personnel to strip firegrounds of all burned material. Now they only remove rekindle hazards, such as mattresses and stuffed furniture, and leave the rest after soaking everything thoroughly. This has resulted in much less work with no increase in rekindles. Some departments have the first due engine company return to the incident periodically to inspect for possible rekindles.

9

EXPOSURE DESIGNATIONS

"A common system of identifying areas within and around a building is needed." - Chief Charlie McKeogh

A good fireground designation system allows everyone on the fireground to identify locations at the scene quickly. Without such a system, many people will use compass designations such as north and south. The problem is that many people don't know north from south especially under emergency conditions or on a dark or curving street. Without a system, long and confusing terms are often used, such as "the building on the east side of the building when facing the front." The term "Exposure B" is shorter, clear and reduces radio traffic.

When you hear the IC make a radio transmission such as "Whatever company is in the easterly exposure, come out now," (an actual message), you can assume that they are not using an effective fireground identification system (or a command chart.)

A Sample Designation System

It is important to use a designation system that everyone (including your mutual aide companies) understands. In the following example each exterior side of the structure has a letter designation.

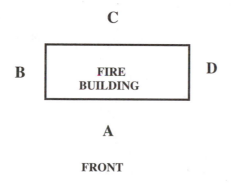

107

Use alphabetic phonetic identifiers such as: Adam, Baker, Charlie and David for designating building quadrants and exposures. Used alone alphabetical letters are easily misunderstood (side C vs. side D.)

In confusing situations, when it is not clear where the front of the building is, side A should be quickly identified by Command to establish a reference point. This also applies to outside incidents such as brush fires.

The buildings next to the fire building are called Exposure B, B/2; B/3; B/4, and so on. The same system applies to Exposure D as in the illustration below.

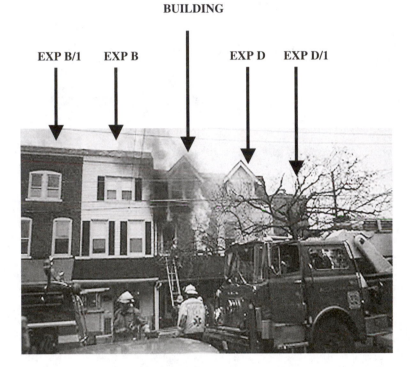

AN EXPOSURE IDENTIFICATION SYSTEM
Courtesy of Dennis Wetherhold Jr.

It is easy to get sides and exposures confused in the heat of battle. The illustration below shows the difference between the two. If smoke is coming from side D of the fire building and is incorrectly reported as coming from exposure D, it can cause unnecessary fireground problems. It is important that everybody on the fireground use and understand the same designation system.

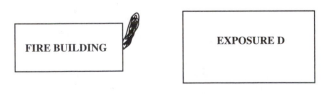

**SMOKE IS COMING FROM SIDE D,
NOT EXPOSURE D**

Locations within the fire building should also be identified by a standard identification system. Knowledge of general locations within the fire building is important to everyone working at the incident. The interior quadrants are A, B, C, and D as shown.

INTERIOR QUADRANTS A, B, C, D

Identifying the location of the fire can improve fireground operations. If the fire is reported to be in quadrant C (right rear), this information helps the attack teams, exposure companies on the floors above, the roof team and Command. Knowing that the fire is located in quadrant C enables the IC to quickly protect Exposure D (the right side.) Without this system, the IC may not know which exposure is the most critical.

Company officers should include quadrants with their radio transmissions.

Examples:

Fire showing top floor quadrant B.

We have a room off in quadrant C, first floor.

We have located several trapped civilians, third floor, quadrant D, need assistance of another company.

When sectors and quadrants are newly adopted, try to "lead" officers so that they can get used to the system. It can be helpful to use both the old and the new systems until personnel are comfortable with the new designations. For example, instead of telling a company to check for fire extension in Quadrant B, and hoping that they remember where it is, you can tell them "check Quadrant B. That will be in the rear of the building, on the left side." This ensures they know where to go and gives them a message that they need to be up to speed on the quadrant system.

When we first started using this system in the 4th Battalion, an officer suggested placing an adhesive diagram of our exposure identification system on all the portable radios. We did it, and it worked great because now anyone in doubt can look at the diagram which shows the exposure and quadrant system.

SUMMARY

A good fireground designation system allows everyone on the fireground to quickly identify locations at the scene.

Use a system to identify the sides of the buildings and the exposures.

Locations within the fire building are identified using interior quadrants.

10

FIREGROUND COMMUNICATIONS

"We have two ears and one mouth so that we can listen twice as much as we speak." - Epictetus

Good fireground communications should be a part of your everyday operation. They are vital for efficient fireground operations. When communications are weak, operations usually go poorly. Most fires that turn into disasters are plagued by poor communications. There is no such thing as good communications at a lousy operation. The following suggestions can improve your department's communications.

BUY THE RIGHT RADIOS

Radios should be "firefighter proof."

Too many fire departments buy "Star Wars" radios— more radio then they really need. Radios should be "firefighter-proof," in the words of an old salty lieutenant. This means that it should be difficult to screw up or make mistakes with the radios. The more features and gadgets a radio has, the more likely you are to have problems operating them. The features of a good practical radio are simplicity, reliability, and ease of operation, especially simplicity. Some radios have tiny knobs and fancy features—like the capability of phoning your stockbroker from the fireground. Needless to say, this function is seldom needed on the fireground!

Besides buying the right radios, the next step is to keep the radios fully operational. Most fire departments buy radios in a group. This means that the radios are all new together, so they also all wear out together. This can result in unreliable radios and batteries. A few years ago, a fire officer died trying to radio for help on a defective radio that had been reported as defective, but had not yet been repaired. Every fire department should have a rule: no reliable radio, no going inside. This means that a fire company does not go inside a building unless it has at least one working radio.

Every maintenance person in every mall in America has a portable radio on their belt to report problems or to be dispatched to clean up a spilled drink in the mall, but we sometimes send firefighters into burning buildings without radios. In many departments, there is only one radio per company! Firefighters without radios must try to stay near the one radio, thus limiting their usefulness and productivity. All firefighters should have a radio for their own safety. Many in the fire service say, "Gee, we can't afford radios for everyone." The truth is that we can't afford not to. How many deaths and near misses must occur before every firefighter has a radio? Some firefighters have radios but leave them behind on the rig because they can't find the radio or they are in a hurry. Firefighters are to be responsible for bringing their radios with them.

ALL FIREFIGHTERS SHOULD HAVE A RADIO

COMMUNICATIONS BASICS

Communications can be complicated, so it is important to focus on the basics.

Arrival Reports

The first company on the scene gives a layout location and then a size-up report—similar to the example below:

Engine 1 (first due) *on the scene of a split level house, ordinary construction, with heavy smoke showing from one window on the second floor, quadrant A (left, front), passing command.*

CONDITIONS UPON ARRIVAL

112

This report indicates the building type and location of the fire in one succinct message. It is an excellent report because it allows everybody responding to visualize the scene. Other responders are no longer flying blind and will be better prepared when they arrive. Besides helping other responders, it also helps the officer who called it in by forcing him to focus on what the problem is. This process can help the officer to make good decisions.

The first company in the rear of a building or incident site should give command a size-up report similar to the one below:

Engine 2 (second due) *on the scene, side C, fire showing from all the second floor windows.*

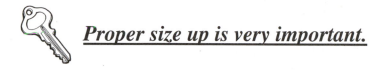 ## *Proper size up is very important.*

Poor Arrival Reports

Often, poor arrival reports, such as the following, are usually screamed over the radio:

Engine 1 on the scene, heavy smoke showing.

All that the other responders know from this report is that there is a fire. The fire could be in a detached residential garage, or in a major building. This report really doesn't help inform the other responding companies. If the first arriving company does not give a complete size-up, the next arriving company should give the report.

Correcting Address

Companies dispatched to a wrong address should give corrected information for the other responding companies. This is common sense, but sometimes companies get excited and simply go into action when they find a fire. Consequently, other responding companies go to the wrong location.

Mask Speak

Good communications are possible while using SCBA. You can effectively talk on the radio through a facemask using a voice amplifier on the mask. The amplifier enhances fireground communication while speaking to one another or while transmitting a radio message while breathing air from the SCBA. Even without voice amplifiers, you can talk and transmit through the mask, however the message will not be as clear.

Report Potential Conditions

The fireground channel should be used to report what you see, what you don't see, and what you think. It is better to report a possible condition (see examples below) immediately than to give a positive report later. It is best to do both.

- *It smells like trash.*

- *There is an odor, possibly electrical.*

- *I think we have an apartment fire.*

Report What You Don't See

If you find no smoke or odors on entering the reported fire floor, notify Command. The actual conditions are obvious to you, but Command doesn't know if you're crawling down a smoky hall, or looking for an apartment number.

Reporting Smoke Conditions

When company officers are checking for smoke conditions in a large building by starting from the top floor and working their way down, conditions on the top floor should be reported as soon as possible. The other floors can be grouped. Officers should not wait until all floors are checked to give a report. For example:

- *Truck 2 to Vent: The top floor number 9 has light smoke.*
  (Always report the floor number of the top floor.)

- *Vent to Command: Floors 6, 7, and 8 are clear.*

Report All Known Or Possible Life Problems

Examples of known or possible life safety threats included in the report are:

- *People are at the windows*

- *People are evacuating the building*

- *People are on the balconies*

Any single report may not be significant, but multiple reports indicate a rescue problem.

REPORT LIFE SAFETY THREATS TO COMMAND

Report All Relevant Information

Don't assume that the IC or others are aware of what you are seeing.

Firefighters arrived to find a store full of smoke, and they thought that the fire was in a storeroom on the first floor. One of the first arriving officers was told by the owner that the fire was in the basement. A firefighter noticed smoke coming up from cracks in the floor tile at the front entrance.

Neither observation was reported or acted upon. It is vital that this type of information be passed to Command. The firefighting forces were unaware that they were operating over a basement fire and the situation turned ugly very fast.

All company officers must understand that the IC is relying on them for information. Danger signs need to be reported. Don't assume that Command or others are aware of what you are seeing. Firefighters can get hurt when everybody on the fireground is not aware of what is happening.

114

Too often crews will operate in areas where conditions may appear normal, while being unaware of critical factors occurring around them. Following a collapse that injured firefighters, it was determined that some officers saw various danger signs that, taken individually, were not critical and were not reported. The combined observations indicated structural problems that ultimately resulted in the collapse. With complete information, Command and Sector Officers can take action to improve firefighter safety.

**REPORT DANGER SIGNS
TO COMMAND**

Communicate Solutions, Not Just Problems
If you need assistance, report the situation and recommend the solution to the problem.

Don't report, *Rescue to Command, we need some more help.* This report is not specific enough and will require Command to have to ask what you need. If you need assistance, report the situation and recommend the solution to the problem, as in the following example.

We have two codes and need another Medic Unit and Engine Company. If you radio only the information *We have two codes.* Command wouldn't know if you're handling it or are expecting Command to supply resources.

Or radio the following when the masks are low: *The masks are running out. We will need a fresh engine company and truck company soon.* Do **not** radio only the following: *The masks are running out.*

Radio the following: *I recommend a second alarm.* Not the following: *We have lots of fire.*

The radio message should be specific and informative:

Attack to Command: We need a backup line.

Vent to Truck 2: We have rescues in front. Open the roof.

115

Take action if possible to resolve problems:

Rescue to Command: We found a hole in the first floor, and have covered it with an old door.

Everybody Listens

On the fireground when everybody listens, the operation will run more smoothly, and fewer messages will need to be repeated.

Listening skills are an important part of communications. Learn to be a good listener. All responding companies should monitor the fireground channel for reports between on scene companies so they will be aware of the fireground situation before arriving. In addition responding companies monitoring the fireground channel can easily be assigned before they arrive on the fireground.

When everybody on the fireground listens, the operation runs more smoothly and fewer messages need to be repeated. For example, if the Exposures Sector calls Command requesting a ladder for checking the attic, Truck 1 on hearing the request, can respond immediately that the ladder is on the way. This action helps Exposures and Command by using heads-up radio work.

Fireground radio communications should be independent of Dispatch. The fastest, safest, and most accurate communications are direct. Avoid relaying radio messages through Dispatch. If you can't communicate directly with your own companies, something is seriously wrong.

The Quiet Radio Problem

The fire was reported to be on the roof, and the response was two engines, a truck company, a heavy rescue squad and a battalion chief. Upon arrival, nothing was showing and companies checked out the roof and the attic while the chief established a command post. I listened to the fireground channel, but there was nothing to hear. The companies checked and went to the command post to inform the chief of their findings. Some officers seem to believe that the less you say on the radio the better. There was no radio traffic at all at this incident. When these companies do have a working fire, communications will probably be a problem from lack of practice. **Every fireground sets the stage for future incidents.**

**COMPANIES MUST
COMMUNICATE REGULARLY**

116

It Sounds Stupid But It Works

Every company becomes an active part of the radio network on every alarm.

As an IC or Sector Officer, do you ever have problems on the fireground because some companies do not answer when you call them? The problem is usually because they are not listening, or they are on the wrong channel. The solution is to talk to all companies on all alarms. Chief Dinosaur says "We are trying to get less traffic on the radio, not more." He is right that we should try to reduce radio traffic at major incidents, by using face-to-face communication and sectoring. But we should increase radio traffic at routine incidents. It may sound stupid, but it works. At routine incidents, find a reason to talk to all the companies, even if it is just to confirm that they received the message to return to the station. This forces company officers to become better listeners and communicators. Using this system routinely helps everyone become accustomed to working and talking with each other so that communications work smoothly.

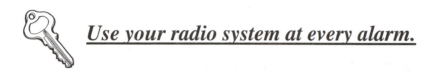 **_Use your radio system at every alarm._**

Acknowledge And Follow Up

If companies fail to get a radio acknowledgment, they usually ignore the radio and freelance.

Radio transmissions should always be acknowledged. Sometimes you hear messages such as: *Engine 2 to Command, we have found the fire on the first floor,* that are never acknowledged by Command. Too often messages are sent but not acknowledged. If important messages are missed, people can die. **Never** allow your radio system to become so clogged, busy, or otherwise messed up that messages cannot be acknowledged.

Assume that messages that are not acknowledged have not been received. If the crews are accustomed to not being acknowledged, the whole system becomes sloppy. If companies fail to get a radio acknowledgment they usually ignore the radio and freelance. Important messages are commonly missed because the IC is not using the ICS and the radio traffic is out of control. Another reason may be that the IC is running around with a portable radio and missing messages.

Too often orders such as: *Engine 8 check for water supply problems,* are given but no follow up report is sent back. When you complete assignments, inform whomever gave you the assignment that your mission was accomplished. When it takes a while to complete an assignment, keep Command informed with regular progress reports. If Command and Sector Officers don't hear back, it's important that they follow up on company assignments to ensure accountability and to make sure that the right things get done promptly.

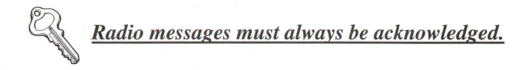 **_Radio messages must always be acknowledged._**

Good Communications Goes Both Ways

Keep your troops informed.

Fireground communication is a two-way street, and it is important for the IC to keep the troops informed about what is happening from the command perspective. Messages from the IC such as: *"The visible fire in your sector looks as if it has been knocked down as seen from the outside,"* gives the troops feedback and another viewpoint. It also gives them confidence that they are part of an organized effort, and that Command is sharp and looking out for their welfare. Acquaint your crews with the game plan. Brief them during long term incidents such as haz mat or police barricade situations, so they know the situation and what is expected of them.

Emergency Traffic

Emergency traffic messages are a very effective way to clear the radio channel for critical messages. Any sector or company officer can communicate with Command when an emergency exists. Most SOPs use the terms "Emergency traffic," " Mayday," "Urgent" or "Priority" to identify critical messages. When Command receives critical messages affecting fireground safety, the IC should make sure that all the troops are aware of the information.

Don't Play Musical Chairs With Radio Channels

The fewer changes made, the less the chance for error.

The fireground channel should be established at the time of dispatch. Companies should arrive with everyone already on the right channel. If a fireground channel is not designated until it is needed on the fireground, communications will usually be poor. This is because once the troops go into action with masks, gloves, and commotion, it is usually a disaster trying to change radio channels. Normally less than half of the radios get the word. The result is chaos, and on the fireground, chaos is dangerous. The fewer changes that are made, the less chance for error. **The moral is: Establish a fireground channel from the beginning.**

Adding Radio Channels To A Working Incident

At major incidents, establishing a command channel for the chief officers will reduce fireground radio traffic. The channel change must be announced and then followed up with a roll call to be sure that the chief officers are on the new channel.

After a command channel is established, the command officers must continue to monitor the fireground channel. A scanning radio is not the answer because a critical message may be missed. Critical messages can come over the fireground or the command channel. The command officers should have a second radio or an aide equipped to monitor both channels.

The best time to add new fireground channels is when new sectors or branches are established, so that existing channels will be unaffected. For example, when an exposures sector is established, assign that sector to a new channel. If that is not practical and it is necessary to have some radios change channels, choose sectors that would be easiest to change. The easiest sectors to change are usually the noncombatant sectors such as water supply or logistics. Avoid changing sectors that are involved in interior fire fighting.

Once additional channels are added, it is very important that the command post continually monitor all the channels. At major operations with multiple channels in use, some radios may be on the wrong channel. There have been incidents when firefighters have called for help on the wrong channel and were never heard. Proper monitoring by Communications Officers addresses this problem.

During major operations, there is always a critical need for more channels. It is wise to have available a cache of radios that are on an entirely different frequency. This doesn't necessarily mean buying more radios. It means knowing from whom extra radios can be borrowed during a disaster and setting up an agreement for their use beforehand. Cellular phones enhance communications and they don't add to radio traffic. Setting up a cellular telephone link between major functions, such as the command post and staging, reduces radio traffic.

**USE CELL PHONES TO
ENHANCE COMMUNICATIONS**

Use Different Channels For Dispatch And Fireground Operations

Some departments use one frequency for both dispatch and fireground operations. Too often, dispatch channel transmissions interfere with the fireground transmissions, killing your communications. Critical messages may be lost or delayed because somebody on the other side of town has a headache and an ambulance is being dispatched. For the same reasons, avoid using radios in the scan mode at emergency operations.

ICS And Communications

ICS improves communications because sectoring reduces radio transmissions, and allows more face-to-face communications. The company officer is responsible for keeping the Sector Officers informed about their location and their progress. Companies should notify their Sector Officer of their progress, when an assignment has been completed, or if they must leave the building because of low air bottles.

Sector officers are responsible for communications from their sector. Before ICS, the IC sometimes received the same message from three different companies, and at other times they received very little information. Sector Officers should give progress reports to Operations or Command at no less than five-minute intervals without being asked. A sample report would be:

Attack to Command: Engine 1 and Engine 2 are attacking the fire, and appear to be making progress.

Who Talks To Whom?

Command often violates the radio chain of command by talking to company officers instead of Sector Officers.

The IC talks to the Sector Officers unless there is an Operations (Ops) Officer to talk to them. The Sector Officers talk to the company officers. This is how the fireground should operate when the ICS and SOPs are used. Unfortunately, that is not always the case. Command often violates the radio chain of command by talking to company officers instead of Sector Officers. This is a very common and serious problem. Even in many "ICS fire departments" the IC talks to companies instead of Sector Officers, or to a mix of companies and sectors.

The reason this happens is that many Incident Commanders fail to sector routine incidents and talk directly to company officers. When an incident escalates, they often continue to talk to company officers instead of Sector Officers. What we do on everyday incidents becomes a habit that is almost impossible to change in the heat of battle.

When the IC talks to company officers on the radio, the command system is starting to fail. Too often the blame is wrongly placed on the ICS system, instead of the fireground commanders.

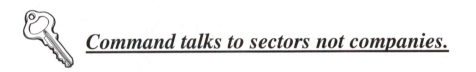 *Command talks to sectors not companies.*

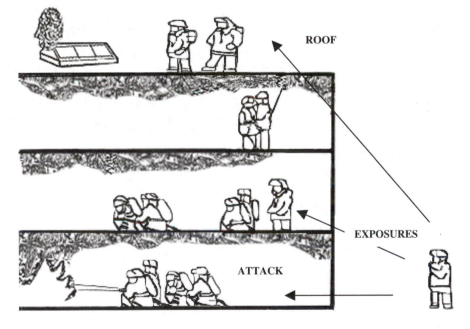

ROOF

EXPOSURES

ATTACK

COMMAND

COMMAND TALKS TO SECTOR LEADERS

The IC Coordinates Communications

Because the IC delegates functions to different sectors, the sectors must be coordinated.

It is critical to firefighter safety for the IC to make sure that the right people hear the messages that apply to them. For example, if the roof team reports heavy smoke coming from the hatch of an exposure, the IC should make sure that the Exposure Sector copied the message. If the Search Sector is operating on the second floor, and the IC gets a report that fire was discovered on the first floor, the IC must immediately notify the Search Sector that the fire is below them. It is important that ventilation and extinguishing operations be coordinated. Engine companies must be notified when truck operations are delayed, and truck companies must be notified of engine company delays. Because the IC assigns functions to different sectors, it is vital that the IC coordinate sector operations.

Messages to the IC are not always accurate. Troops operating inside fire buildings usually have tunnel vision. Most fire departments have tales of inside companies reporting that the fire is under control when in fact it is coming through the roof. Interior reports are often best estimates coming from hot, smoky, and often confusing situations. The IC must keep this in mind when receiving and coordinating fireground communications.

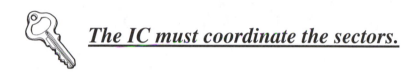 **_The IC must coordinate the sectors._**

Incident Progress Reports

Command should give progress reports at regular intervals. The purpose of the reports is to advise Dispatch and responding companies what the situation is. The IC also benefits because formulating a periodic progress report forces the IC to stop for a minute and think about exactly what the situation is at that moment.

The Communications Officer

At major incidents it is wise for the IC to assign a Communications Officer. This officer helps to keep communications operating properly by:

- Recommending the establishment of new radio channels when needed

- Working with the phone company to establish telephone lines

- Setting up fax and cellular telephone links

- Ensuring that all the channels are monitored for safety at each incident

- Working with mutual-aid fire departments to handle radio frequency problems

It is important to ensure that your radio equipment (and procedures) are adequate and sufficient to support radio traffic at multiple responder firegrounds.

DON'T WASTE RADIO TIME WITH POOR MESSAGES

All radios should have two stickers attached:

LEARN BY LISTENING	**PUSH TRANSMIT BUTTON WITH BRAIN NOT THUMB**

Radio discipline is critical because we can't afford to waste radio time. Too many incidents have messages such as the following:

Exposures to Command, we're over here in one of the exposures. It used to be a shoe store, but I don't think it is any more, and I'll let you know when we find something out.

After the fire we can laugh at radio transmissions like that, but they eat up valuable radio time.

Fireground communications should be complete, concise, and clear. The following examples of poor transmissions were taken word for word from the radio tape:

Engine 1 to Command, it's coming out the window. (What's coming out the window: fire, smoke, water, confetti?)

Command to Truck 14, go to the rear and see if you can help them there. (Help whom, with what?)

Command to Engine 28, take the second floor. (And do what: fire attack, protect exposures, rescue?)

Roof to Command, the roof hatch has been taken off. (What was the result? Is smoke, fire or nothing coming out?)

The following radio traffic shows why many IC's get prematurely gray.

Command to Squad 3, is the secondary search complete? (The IC shouldn't have to ask.)

Squad 3 to Command: We haven't found anybody, and we are opening the ceilings. (Are they looking for people in the ceiling?)

Command to Squad 3: Does that mean that the secondary search is completed?
Squad 3, Affirmative.

The moral is that fireground communications should be complete, concise, and clear.

Terminology Problems

Many fire departments believe that terminology problems are between fire departments, not within a department. Actually, it is common to have terminology problems within a department.

Problem Cases

A classic case was captured on audio tape during a working house fire. The tape begins at the point where the IC wanted a report on conditions in the rear of the structure, and called a lieutenant in the rear for a report. This is how it went:

Command to Engine 2: How is it looking back there?

Engine 2 to Command: It's going good.

Command to Engine 2: Is it going good-bad or is it going good-good?

Engine 2 to Command: It's going good bad. (This really happened. You can't make up stuff like this.)

I was a Lieutenant at a multiple alarm fire and stumbled upon an exposure that was so hot that it was starting to smoke from the radiant heat. I called the IC and told him that we had a severe exposure problem. He thought I said we had a severe explosives problem. I didn't think the old man could run that fast.

Some communications become mixed up because there are no SOPs. One department uses the term "Take your assigned positions," but the term is not written in the SOPs. Most members think it means to hold up or stage in line of approach, but it can be interpreted different ways.

At another incident, Command wanted a second alarm. This is what happened:

Command to Dispatch: I need a second alarm.

Dispatch: Are you asking for a second alarm or a second line?

Command: That's correct.

Dispatch: O.K.

It was ten minutes before this snafu was corrected and the second alarm sent. **Make sure that everyone is talking the same language.**

Poor Fireground Terminology

One afternoon I was the IC at a major fire in the downtown business district. The fire was threatening to come through the roof. The exposure building was attached and much higher than the fire building. This situation made it a potentially severe exposure hazard.

To protect the exposures, I ordered the truck company to "set up their ladder pipe," intending that it be ready for action if it was necessary to protect the exposures. When I told them to set up the ladder pipe, I could tell they were not very happy with the order, but I had too many other things happening to pay it any mind. It turned out that the fire was stopped on the top floor, so the exposures never really became a problem. In talking with the truck folks later, I found out that they thought I wanted them to start using the ladder pipe in the middle of an interior operation. I thought "set up" meant get ready to do it, but the truck thought that "set up" meant do it. This incident shows the importance of having common terminology. It ended well, but in a worst case scenario, we could have had troops on the top floor engaged in an interior attack and the truck company opening up with master streams from above.

Another example of problems with a lack of common terminology involved an engine company that was directed to cover the rear (Exposure C.) The company soon reported back that there was no exposure problem and requested another assignment. They were directed by Command to "cover your position" — not a defined term in that department. It seemed that the IC was busy and not sure of what to do with them. What they did was freelance.

123

Still another example occurred in a fire department that had an evolution in which a standpipe pack was carried up into any large buildings that were not equipped with standpipes. When the fire was located, the standpipe pack was brought to the nearest window and the end of the hose lowered to the ground. Through the years, the evolution became known simply as "dropping the pack out the window." One night, at a fire, an officer ordered a firefighter to "drop the pack out the window" The excited firefighter did what he was told. To the horror of the officer, the firefighter dropped the whole pack out the window! Using the right terminology is important. From then on, the firefighter has been known as "Drop Pack Williams." (Name changed to protect the guilty.)

Ideal Working Fire Radio Sequence

Fireground radio communications should flow. This means that there is a smooth, reliable, and predictable stream of messages at every incident. The following radio transmissions offer an example of good fireground communications. Although not printed here, all transmissions are acknowledged by the recipient.

Engine 1 dropped a supply line from 3rd and Oak Streets.

Engine 1 on the scene of a two-story brick townhouse, ordinary construction, with heavy smoke showing from the second floor, Quadrant A (left, front*), passing Command.*

Engine 3 providing water supply to Engine 1 and assuming Command.

Engine 3 backing up Engine 1 in the Attack Sector.

Engine 2 on the scene in the rear with fire showing 2nd floor, quadrant B (left, rear.)

Engine 2 to Command: The basement is clear, and we're protecting Exposure B (*building on the left), and assuming Command of the Exposures Sector.*

Truck 1 on scene. Assuming the Vent Sector, and opening roof hatch.

Battalion 1 assuming Command.

Command to Engine 4: Cover Exposure D (right side.) *You are assigned to the Exposures Sector, did you copy this, Exposures?*

Exposures copied.

Exposures to Command: The 1st floor of exposure B is clear. We are covering the second floor. Light smoke.

Roof team to Command: The hatch is open, heavy smoke is coming out. Both sides look okay from here.

Rescue to Command: Primary search complete. Negative.

Vent to Truck 2: We are working the fire floor. Check below for salvage operations.

T-2 to Vent: We're throwing covers on the first floor.

E-4 to Exposures: The cockloft of exposure D has heavy smoke but no heat.

Exposures: Command copied E-4's message.

Attack to Command: Fire appears knocked down. Checking the overhead.

Rescue to Command: Secondary search complete. Results negative.

Vent to Command: Electricity has been shut off.

Exposures to Command: Smoke in exposure D is clearing up.

Command to all companies: Air monitoring negative. All clear, Okay to take masks off.

T-2 to Vent: First floor's covered. Basement okay.

Vent to Command: No fire extension.

Exposures to Command: Exposures B and D clear.

Command to Attack and Vent: Go to rehab for a break. Engine 2: You are now assigned to the pre-overhaul safety check of the fire building.

Another happy ending. No muss. No fuss. No errors and no screaming or lost messages. All the bases were covered. The attack crew attacked, and the backup team protected them. The building was vented and searched, and the exposures were protected. Things went well for the IC because this fire was fought using SOPs and management-by-exception principals.

Talk To Your Neighbors

Mutual aid is the only way that most fire departments can get enough resources to handle major incidents, so make sure that your radios are compatible with the radios in neighboring departments. Establishing good working relationships with mutual-aid departments before you need their services is critical to a smooth running radio system. Everybody agrees with this, but few do it. It takes a lot of foresight and work to make it happen.

When mutual-aid companies are on separate radio channels, it is not necessarily bad. Sometimes these companies can be grouped together using their own frequency during major operations. The Sector Officer is then assigned a radio from the host department.

SUMMARY

If you use good fireground communications at routine operations, you will have good communications at working incidents.

Radios should be "firefighter proof," meaning that the radios should be difficult to screw up. Avoid those with too many bells and whistles.

The initial arrival report should indicate the building type and the location of the fire in one succinct message. This allows others who are responding to be better prepared when they arrive.

The fireground channel should be established at the time of dispatch.

Fireground communications should be complete, concise, and clear. You need to report what you see, what you don't see, and what you think.

All company officers must understand that the IC is relying on them for information. Danger signs need to be reported. Don't assume that the IC or others are aware of what you are seeing.

Radio transmissions must always be acknowledged. Messages that have not been acknowledged should be considered not received.

On routine incidents, Command should try to talk to all companies, so everyone becomes accustomed to working and talking together, so that communications will work smoothly at working incidents.

Use of emergency traffic messages is a very effective way to clear the radio channel for critical messages, such as firefighters in trouble or reports of serious situations.

ICS improves communications because grouping companies into sectors reduces radio traffic. Sector Officers give progress reports at least every five minutes, without being asked.

When sectors and companies need assistance, they should report the situation, and recommend the solution to the problem. Avoid simply reporting problems.

Because the IC delegates functions to different sectors, it is vital that the IC coordinate sector operations. The IC must make sure that the right people hear the messages that apply to them.

Fireground communications are a two-way street and it is important for the IC to keep the troops informed what is happening from the command perspective.

At major incidents, establishing a command channel for the chief officers reduces fireground radio traffic.

The best time to add new fireground channels is as new sectors or branches are established, so that existing channels are unaffected.

Once additional channels are added, it is very important for the command post to continually monitor all the channels. During major operations, a Communications Officer should be assigned to help keep communications operating properly.

Cellular phones provide good communications and reduce radio traffic.

Make sure that your communications equipment is compatible with your neighboring departments.

11

SAFETY

"Take calculated risks. That is quite different from being rash."- George S. Patton

In the fire service today, there is a lot of controversy over safety. Our safety record has been poor, and partly because of this, many regulatory agencies and groups are forcing regulations down our throats, supposedly to make our job safer. We are being over regulated, and we don't like it. Too many of these requirements are needless or impractical, focusing on fixed rules with a one-size-fits-all approach. Many people in the fire service resent meddling from folks who can't tell a nozzle from a ladder. Some proposals seem to be trying to make a totally safe fireground, which we all know is not practical.

On the other hand, there are some very real safety problems that are routinely ignored by both the regulators and the fire service leaders. Our safety efforts tend to focus on hardware such as helmet standards, but we need to focus more on what is inside a firefighter's head than what is on top of it. Most of our typical injuries are caused by human error, poor judgment and just plain dumb actions. Most of our safety solutions involves making more regulations and buying more expensive safety equipment.

Safety statistics show that for every serious disabling injury that occurs, there are 10 minor injuries and 600 near misses. This safety information is a well-kept secret in the fire service. Very few are aware of these probabilities and even less do anything about it. Often minor injuries are not seriously reviewed "because nobody really got hurt." Reducing the causes of the minor injuries will reduce the rate of serious accidents. Near misses are difficult to track, but patterns or serious near misses should be reviewed. The folks reviewing safety should include some front line fire people, not just the paper pushers.

SAFETY PHILOSOPHY

Is Safety Really # 1?

What kind of safety record does your fire department have? Are your injury rates up or down? How are you doing nationally compared to similar fire departments? Unfortunately, most departments really don't know. Even if the Safety Officer and the brass are aware of the statistics, few of the rank and file have a clue about injury rates, and, moreover, many just don't care.

We all keep injury and accident statistics, but sometimes we forget that we should be using the information to reduce injuries, not just document them. The stats should be used to determine the safety problems, come up with solutions, make changes to reduce injuries, and check to see if the changes made any difference. If your department is doing this, you are in the fortunate minority. Although we complain about unnecessary outside regulations, we should also realize that we need to be doing more to improve safety in our departments. This chapter will give you some insight, inspiration and ideas for improving safety in your fire department.

The sad truth is that what we say about firefighter safety and what we do about it are often two different things. Consider a fully involved abandoned building with no exposures. The only important consideration at this fire is firefighter safety. Besides, everybody will probably be better off if the place burns down. Extinguishment and water supply are secondary. While safety is the only thing that is really important, very few fire departments will use a Safety Officer at this fire. Typically, it's Firefighters 15, Safety Officer zip. Usually all the firefighters are committed to extinguishment or water supply, while we insist that safety is our first priority. Every fire department needs to evaluate their safety practices to narrow the gap between what they say and what they actually do.

Several firefighters have been lost and subsequently died because their fire department didn't have the resources to provide PASS devices. Guess what! Now all their firefighters will all be equipped with PASS devices.

**TOO OFTEN PROGRESS IN THE FIRE SERVICE IS
BASED ON TOMBSTONES**

 ___*Focus on safety, and make changes to reduce injuries.*___

Enforcing Safety Operations
One of the reasons why we are often lax about enforcement is that we don't take safety seriously.

We have rules and disciplinary systems, but we seldom do anything about safety infractions. Miss a meeting, or show up late and you will likely wind up in trouble. However if you go

128

into a burning building without all your gear, or fail to activate your PASS device, it can be ignored. This tells us something about our priorities. One of the reasons why we are often lax about enforcement is that we do not take safety seriously. If safety violations occur, they should be followed up.

Too often Lieutenant Kamikaze has a history of being a cowboy, but nothing is done about it. Sometimes the safety infraction becomes known after the lieutenant has been hurt, and there is a reluctance to go after somebody who is hurt. Often, the violations are made by the aggressive go-getter firefighters who are respected and valuable, and that can cause some officers to look the other way.

In some fire departments, more emphasis is placed on fire station safety than on fireground safety. Wipe up the floor and don't leave the mop bucket in the hallway, but when it comes to the fireground, safety goes out the window.

Safety can't be turned on and off like a light switch. Safety must be part of our everyday operations. When safety practices and rules are regularly ignored, we are setting the stage for disaster. At dangerous and complex incidents or when routine incidents turn sour, you can't switch on safety. If a crew that regularly investigates odors of smoke without being fully prepared finds a working fire, it is unlikely that they will suddenly operate safely.

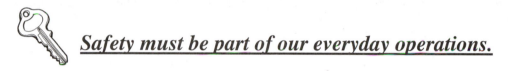

Safety must be part of our everyday operations.

Lessons From Europe And Industry
If the brass pretends to be interested in safety, the firefighters will pretend to practice safety.

Firefighter safety is given a much higher priority in European fire departments than in this country. They will not risk firefighters' lives to save property, and they are much stricter about using masks. In many European countries, fire officer's are penalized for not following established safety procedures. It makes sense that if we consider safety really important, people will be disciplined for failures to follow safety rules.

Effective industrial safety programs show that commitment equals success. Companies that have outstanding performance records invariably have made a full scale commitment to safety. How may fire departments can claim a 100% commitment? If firefighters know that their department is really concerned about safety, they are more likely to work safely. You can't fool the troops—if the brass pretends to be interested in safety, the firefighters usually pretend to practice safety.

Positive Safety Incentives

In fire departments, it can be easier to discipline than it is to reward. Most of our safety incentives are negative. If you don't follow the safety rules, something bad may happen to you. Industries often post signs that read, "NO LOST TIME INJURIES AT THIS PLANT IN 101 DAYS." Have you ever seen a sign like that in a firehouse? I haven't either. We should reward those who practice safety and play by the rules. We should be publicizing good safety records and rewarding drivers for their safe driving records. Award companies and individuals that go for a certain period without an accident or injury. Pins, patches, and awards are an inexpensive and practical way to get the safety message across.

REWARD EMPLOYEES WHO PRACTICE SAFETY
Courtesy of Sun Gazette Newspapers

Cardiac Alert

More than one third of all firefighter deaths are caused by heart attacks. About 90% of those dying from heart attacks had a history of cardiovascular problems. It doesn't take a medical genius to figure out what the problem is, or how to prevent it. The obvious answer to this important problem is to conduct complete medical evaluations regularly. If you can't meet the standards, you can't fight fire. We also need to encourage fitness and healthy lifestyles.

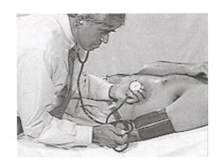

REGULAR PHYSICALS ARE VERY IMPORTANT
Courtesy of Loyola University Medical Network

SAFETY HORROR STORIES

These incidents say a lot about our lack of safety consciousness.

Consider this statement from a national fire publication. "The truck officer decided not to risk ventilation of a large unsupported roof during a church fire, so they went *inside* the church (emphasis added) to assist the engine companies with an interior hand line attack." If it is too dangerous to be on the roof because the stability of the roof is in question, it's crazy to join forces with firefighters who *are operating under the roof.* Even if a poor decision was made in the heat of battle, it's inconceivable to write about your error, as if it was the right thing to do! This incident says a lot about our mentality and lack of safety consciousness. So do the next stories below.

The Night The PBI Running Pants Burned Up

The cover-up was sent in on the reports and was never questioned by the higher ups.

The TV news flash stated that a firefighter was badly burned that afternoon in a store fire. The fire was in the basement. The firefighter had partially slipped through a hole in the floor and became trapped and burned. The next time I saw somebody from that department I asked them how this happened. I was told (by someone with a straight face) that the firefighter's PBI running pants had burned off, exposing him to the flames. I then asked what *really* happened. After some gentle persuasion, the story came out. The fire company was returning from a medical call and got the response as an odor of smoke. The firefighter (and maybe others) didn't take the call seriously, and failed to put on his running pants. He fell through the floor and was seriously burned. The cover-up was sent in on the reports, and was never questioned by the higher ups. There was no investigation, and nothing changed. The department is back to business as usual.

The Day The Pipe Man Almost Died

The pipe man came tumbling down the stairs on fire.

The call was for a house fire with a report of a person trapped. The first engine arrived with a crew of three and found smoke showing from the second floor. The truck company was almost there, since it came from the same station. Upon arrival, the pipe man ran into the building without a line, apparently to see if anyone was trapped. When the officer and driver started in the front door with a hose line, the pipe man came tumbling down the stairs on fire. They charged the line and used it to extinguish their pipe man who was burned so seriously that he was disabled and was never able to return to work.

The aftermath was interesting. Some people in the department recognized it as a screw-up. Others believed that since the firefighter was injured attempting a rescue, he must be a hero. Some thought, that the incident represented just another hazard of the job. The fire department never came to an official conclusion. The same thing could occur today. Nobody learned anything from this unfortunate incident.

Playing The Slots

Firefighter injuries and deaths are like the deaths from the Titanic sinking. People didn't die just because the ship hit an iceberg. There were many contributing factors. If just one of these events had not occurred, or occurred in a different order, the losses would not have happened. Firefighters are rarely killed by a single cause. It is almost always a chain of events. For example, at a "routine" house fire where three firefighters became trapped and died, the investigation disclosed more than a dozen deficiencies in the fireground operations, ranging from poor command to lack of accountability. It is reasonable to conclude that none of the problems was new; it was just the first time that they all occurred at the same fire and in the wrong order. The result was a tragic death toll.

Every fireground and emergency scene has similarities to a slot machine. At each incident, you pull the handle and hope that all the lemons don't line up, because that means that something bad happens. The lemons are predictable. The lemons represent problems with: poor accountability, communications, command, control, ICS, sectoring, strategy and tactics, company integrity, and freelancing. If your incidents aren't normally sectored and companies don't routinely stick together, there are already two lemons showing before you even start. All it takes is a few more lemons to line up for something really bad to happen. On the other hand if you're are doing everything properly on a day to day basis, the odds are strongly in your favor.

DEVELOPING SMART AGGRESSIVENESS

Fire fighting is dangerous. The safest way to fight a fire is to stand outside, out of the collapse zone and throw water. That is a safe operation—usually not practical, but safe. While safety is very important, there is a balance between controlled risks and total safety. Sometimes you have to take chances to save lives. Knowing when to be aggressive and when to hang back is the key. There is a fine line between being aggressive and being stupid. Street smart firefighters use controlled aggression, by weighing the potential gain against the loss.

.

Before taking action at an emergency scene, and periodically throughout the incident, we need to ask ourselves:

- What do we stand to gain?

- What can we lose?

If the question is whether to make an interior attack on a heavily involved abandoned building, the potential gains and losses are:

- We may save part of an abandoned building.

- We could easily have firefighters killed or injured.

The decision at this fire is a no brainer. The risks are great, and the gain is nil. Putting safety first and going defensive is the only reasonable option.

A smart risk policy (All firefighters should be reminded daily of this policy.)

- We will take a reasonable risk to save life.

- We will take minimal risk to save property.

- We will take no risk for what is already lost.

Too often our philosophy has been to risk our lives to save property, sometimes at the cost of our lives. Fortunately, we have become more safety conscious in recent years, but we still take too many chances. In some cases firefighters should make their own risk-benefit judgments regarding life safety. Personnel should question orders that appear to be too risky. Supervisors must be trained to accept these questions and make decisions that ensure firefighter safety. **No building is worth a firefighter's life.**

Always consider the risk against the gain.

Dumb Aggressiveness

There are times to be cautious, and times to be aggressive. The wise firefighter can tell the difference.

Firefighters are aggressive by nature, which is good when it comes to saving lives. The problem is that too often we are aggressive in the wrong situations. Being aggressive at a haz mat incident is dangerous. Being aggressive at an abandoned building fire or at a building that is heavily involved in fire is dangerous and dumb. There are times to be cautious, and times to be aggressive, and the wise firefighter can tell the difference. Some firefighters believe that we are always supposed to take action, but sometimes the best option is to maintain a control zone, or wait for the power company. Sometimes firefighters have problems hanging back, so they do dumb things just to be doing something. **Doing nothing is better than doing something stupid.**

The Haz Mat Approach To Structure Fires

Fire departments had to change gears when we began to handle hazardous materials incidents. The slam bang full-speed ahead approach that we use for structure fires didn't work with haz mat. We've adjusted and we now approach hazardous materials incidents with a methodical "lets talk about this, before we take action approach." Perhaps it's time to start using more of the haz mat approach at fires. I know, I know, I'm tampering with sacred beliefs. Tradition stands squarely in the way.

Typically when we see fire, the bugle plays charge and the risk versus the gain principles go out the window. We fly into action charging in, throwing water, searching, venting and all the usual. This is our big fire and we are going to kill it. It's macho, and it's also an easy way to get hurt. What we should be doing is stopping for a few seconds to think about the situation, especially if it is not your typical fire.

Size-up is not just a command function, firefighters should do a fast size-up before taking action. For example, if the fire is in a warehouse, we need to be thinking that we are facing a very dangerous situation. Forget charging blindly ahead, consider using an exterior defensive attack from the beginning, or if it is reasonable to go inside, use search ropes so you can find your way out. If the fire is in a boarded up building, we need to be thinking: lets get the boards off and vent the place before we go inside, if we go inside. If the fire is in a commercial attic or cockloft, we should be thinking collapse. We need to take the time to use common sense and think more about what we are doing and why we are doing it.

"That all sounds good" gripes one officer, "But the reality is that if I paused to evaluate the situation, the next company would be pushing us out of the way to rush in. I would be doing the right thing, but they would get the fire, and I would look like an idiot." He has a good point. To make this work you need strong command, sectoring and a supportive department philosophy.

WHAT THE SAFETY OFFICER DOES

Have you ever looked at a fire as a bystander? If you have, you have probably noticed many things that you wouldn't notice if you were involved with fighting the fire. You get a different perspective when you aren't doing the emergency work. The Safety Officer can focus in on scene safety like a laser beam.

The wise IC designates a Safety Officer to improve scene safety. If the IC doesn't assign a Safety Officer, the IC is also the Safety Officer. Wearing two hats during a working incident is difficult at best, fatal at worst.

The Safety Officer should be someone who is knowledgeable and experienced. In some departments, the last person to show up gets the safety vest, because it was the last one left.

In other departments, the Safety Officer is a paper pusher for the fire chief who doesn't have a clue on strategy, tactics, or fire fighting. According to one officer, "In my department the Safety Officer got the job by being evicted from the front office. He never led a company, he is a joke and is ignored." Choose a Safety Officer who is competent in fire tactics, credible among the troops, and knowledgeable about safety.

The Safety Officer should be a part of the command system, not just a safety cop. This means that the Safety Officer's job is not just to catch violations, but to monitor a system where safety is already built into the system. If the system is working properly, the Safety Officer shouldn't have to tell people to use their gloves or to keep on their helmets. These things should be a part of an established routine. The Safety Officer concentrates on the big picture. Are conditions in the rear consistent with an interior attack? Are inside forces hampered by lack of ventilation? Is the building stable? The Safety Officer makes safety recommendations to the IC and, in a critical situation, has the authority to stop an unsafe operation. Because the Safety Officer is normally moving around watching the entire operation unfold, this officer is often in an ideal position to send reconnaissance information to the IC.

 **_Use a Safety Officer at working incidents._**

Sure We Want A Safety Officer, But...

Personnel shortages should not be used as an excuse for not having a Safety Officer.

We all agree that working incidents need a Safety Officer, but most of these incidents don't have one. The reason is usually attributed to a lack of sufficient personnel. Chief Dinosaur says "Hell, I don't have enough people to put the fire out, and now they want me to appoint a Safety Officer." Certainly personnel availability is a consideration, but there are ways to have a Safety Officer if you really consider safety a priority item. Personnel shortages should not be used as an excuse. Volunteer departments usually have some senior members who are experienced but not up to the physical demands of front line fire fighting. They would make good Safety Officers. Some departments solve the problem by automatically dispatching another chief officer to working incidents as the Safety Officer.

FIREGROUND ACCOUNTABILITY, A LIFESAVER

No one knew that the two firefighters had re-entered the building. They were not missed until their bodies were found during overhaul. Accountability is a difficult but critical process. It is relatively new to the fire service, and few fire departments have accountability systems that really work well.

> "Too many people in my department don't take accountability seriously. I often see missing tags, firefighters without tags or in someone else's gear with the other guy's tags." (Volunteer Lieutenant Jim Holian)

"In my department, accountability is a joke," complains another firefighter "The only time that we ever do accountability is when the fire is out. We never do it when things are getting really bad, which of course is when you really need it."

Accountability saves lives by tracking companies and firefighters. Your accountability system must be simple and accurate. It has to be used regularly, not left on the shelf until a major incident occurs. Some departments start accountability when a second alarm is sounded. What about the accountability of the first alarm companies at a working fire? Playing catch up with safety is never a good idea.

Accountability systems are designed to prevent firefighters from becoming missing, not to simply identify missing firefighters. Accountability reduces freelancing. There is little wiggle room for the kamikaze firefighters to do their own thing because every firefighter and company are always accounted for. If the freelancer joins the wrong company, the company count will be plus one, indicating a problem. If the freelancer is inside operating alone, he should be quickly spotted and challenged by Sector Officers, because no one should be working alone. Keeping people together at their assigned locations avoids problems and reduces the chances that firefighters will be lost.

In order for your accountability system to be effective, it must always be able to give you feedback to four critical questions. These questions are:

- Who is at the incident?

- Where they are at the incident?

- What are they doing at any given time?

- What are the conditions?

You can check your accountability system by asking yourself these questions during your next fire.

Accountability systems reduce freelancing.

Stay Together And Live

At a working basement fire, a firefighter told his officer that he was going out for a new air bottle. The firefighter became lost and ran out of air. He was found in time, but it was a very close call. Recently, a probationary firefighter became separated from his crew and died. The firefighter was low on air and attempting to exit a commercial building alone, when he became lost. Companies normally enter together, so if one bottle is low, they should all be low. It is dangerous for a firefighter who is low on air or having problems, to attempt to exit alone. Too often the lone firefighter fails to make it back out. Nationally about 25% of firefighters who die inside structures, perish by becoming lost and dying of asphyxiation.

It is essential that officers keep their companies intact. This is called company integrity. Each crew should enter together, work together and leave together. Whenever one member of a company or team has to leave, the entire team must leave together for safety reasons. In many departments it doesn't work this way. Too often the emphasis has been on remaining with the hose line as long as possible. The thinking (or lack of thinking) seems to be the longer you can stay, the better a firefighter you are. Little consideration is given to the safety of the hose line crews.

The minimum crew size in the hazard area is two firefighters equipped with portable radios. All personnel should be in contact with their officer by either voice, touch, or sight. This is easy to say and hard to do, but it improves safety and simplifies the accountability system.

When crews stick together on routine smells and bells calls, they will remain together at the workers. And, if conditions go to hell, the crews will be together and will exit as a team because that's what they always do.

THE CREW THAT STICKS TOGETHER GETS OUT TOGETHER

Companies must enter, exit, and work as a team.

An Accountability System

Your system should be the same or compatible with those used by your mutual aid department's. Most systems use Velcro or clips to attach personalized name tags to the firefighter's turnout coat. Typically, firefighters place their tags on an accountability board on their apparatus, and on the small board attached to the officer's portable radio. Apparatus operators who will normally be working outside the structure are identified on both boards. When the company is assigned, the officer leaves the small board either at the command post or with the accountability officer, depending on the circumstances. The apparatus board remains as a backup to the system.

The small boards that are collected are placed on the command accountability board. The boards are organized by sectors, showing each company type and number and the individual members of the company. If a company becomes missing, or is missing personnel, the boards are used to quickly identify the missing members. If the boards have not been collected, they usually remain on the apparatus. That means that if a firefighter is missing, someone has to go around the fireground collecting and organizing the boards, a maddeningly slow process when you may have missing firefighters.

**ACCOUNTABILITY IMPROVES
FIREFIGHTER SAFETY**

*Courtesy of Fairfax County Fire and Rescue,
Fairfax, Virginia*

The Truth About Accountability

The truth is that accountability tags are just trinkets. Accountability tags do not make you safe. They are tracking tools that do not take the place of staying together and using good judgment. People and their attitudes make accountability work. Your real accountability system is in your day to day procedures of crew integrity, sectoring, following SOPs and operating within the ICS system. The formal accountability system is a double check to ensure that your system is working.

You might think that after you assign an Accountability Officer you no longer have to be concerned with accountability. WRONG! Accountability is not a one person job; it is a team effort. Firefighters are responsible for sticking with their company. Company officers are responsible for keeping their company together. Sector Officers keep track of the companies under their command. The Accountability Officer keeps track of everybody. The IC has overall responsibility for accountability. Together, all personnel work together as an accountability team to ensure fireground safety.

**Your real accountability system is staying together and following SOPs.**

Accountability Checks

When accountability systems first began, many departments scrambled to put together a system to comply. However, after the procedures were adopted, many departments never followed up and evaluated the system to make sure that the bugs were worked out. One department required the first accountability check 30 minutes into the incident. Nobody became familiar with the system because few incidents last that long and those that do are usually winding down. The result was that hardly anybody had a chance to use the system, and when it was needed few were familiar with it. Then a fire occurred where the accountability system failed miserably (and predictability), and it took forever to account for missing firefighters. After this fire, the only change was that the accountability checks were required after twenty minutes, instead of thirty. The only thing that really changed was the rule book. Another year, another fire, and the same thing happened again. Now they hold ten minute checks, and their accountability system works reasonably well. Additional checks are made at ten minute intervals. It is important to make sure that your accountability system really works; you don't want to discover a problem only when the system fails.

"That's a terrible time to do it," roars Chief Dinosaur. "Ten minute checks are ridiculous. That's too soon and too often. We are setting up and in the middle of everything. It will slow things down and mess up everything."

When we are in the middle of "everything" is when we lose firefighters. Accountability checks prevent firefighter deaths and must be used when they are needed most. Twenty or more minutes is more than enough time to get lost, run out of air and die many times over. Chief Dinosaur's experience is that accountability is a slow process because he does it so seldom that he is not very good at it.

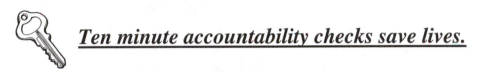

Ten minute accountability checks save lives.

In addition to ten minute intervals, accountability should be checked at incident milestones, such as:

- When an operation changes from offensive to defensive

- When the fire is controlled

- Whenever firefighters are injured, in trouble, or reported missing

Quick Accountability

Chief Dinosaur says: "The radio is too busy to interrupt for a roll call of all the companies." Question: What messages are more important than firefighter safety? Practiced efficiently and regularly, accountability can be done very quickly. It can become a thirty second process, instead of a five-minute event. Accountability is slow in some departments where the IC announces a roll call that stops all radio traffic. In those departments where the IC personally calls each company on the fireground it's no wonder that accountability checks are done infrequently if at all. Command should not have to check all the companies; that's the job of the Sector Officers. Command (or the Accountability Officer) need only to talk to the Sector Officers who check on the crews assigned to them.

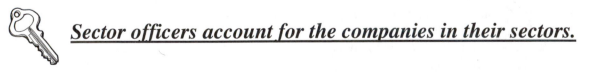

Sector officers account for the companies in their sectors.

To speed up accountability, require Sector Officers to include an accountability check along with periodic situation reports when possible. An example is: _Vent to Command: the roof is opened and we are PAR_ (all accounted for.) Command can check-off that Vent has accounted for everyone, noting the time. If these messages are routinely given by Sector Officers, a complete accountability check requires checking only with those sectors who haven't reported PAR within the last ten minutes. This spreads out the accountability checks so that they are not necessarily all done at once.

Accountability Problems

Even when you use accountability regularly, making accountability work at large incidents can be a challenge. It is easiest to keep track of companies that respond together from the station, but that is not the reality in many fire departments. In many volunteer departments, firefighters respond from home. In career departments when fires occur near the change of platoons, oncoming crews or individuals arrive and sometimes just go to work. This is also true of members called back from home for fireground duty. Career firefighters may also show up on firegrounds because they belong to fire paging systems that alert them to working fires. Too often, arriving firefighters jump right in or report to an overworked IC who just tells them to "help out."

All of these situations can create accountability nightmares. The answer to this problem is to have clear reporting-in procedures, strong command and a staging area for arriving members to report to. All arriving members must be assigned to a company or team for accountability. Some departments hold reporting members back until they can assign personnel in pairs. Others wait until there are enough personnel for a new company.

When firefighters are in trouble, there is often confusion about exactly who made it out and who is still inside. Too often wrong assumptions are made. Guesswork doesn't hack it when we are talking about missing firefighters. When bad things are happening on the fireground, it is time to be extra cautious. At one fireground, firefighters exiting the building were mistaken for the missing crew, and the search for missing firefighters was stopped. The firefighters who were still trapped died on the scene. Strict accountability can be a lifesaver in these situations.

During the early stages of an incident some truckies may be doing outside ventilation, while the other members of the same company are inside the structure. Likewise, engine companies typically have a pump operator on the outside working the pumps. How can an inside company or Sector Officer account for the outside firefighters? How can an officer account for the outside firefighters if they decide to go inside without telling anybody and not joining their crew? The best answer is for company officers to make it clear to their outside members that they must either remain outside or rejoin the company, and that freelancing will not be tolerated.

When your accountability checks always show that everybody always has all their crew with them, be suspicious. "I have witnessed officers working by themselves who said that they were PAR when they had no idea where their crew was" (Firefighter Freddie Schall.) Make sure that your officers are not just playing along with the system to keep the chief happy. Firegrounds are confusing places, and there will be times when officers will need a little more time to check everyone's welfare.

Accountability doesn't work very well in many departments. Probably the most common problem is that many departments have an accountability system on paper that is not really followed. Firefighters get lazy about placing their tags the accountability board. Company officers often bypass the system because they are in a hurry to do battle, so they leave their board on their apparatus. Too many chief officers do not check accountability because they are busy or are not taking the incident seriously. When incidents are not properly sectored, accountability is difficult.

Making Accountability Work

Have a commitment from the top and use the system regularly.

At working incidents, the IC should designate an officer to handle accountability if possible. The IC is responsible for accountability if an Accountability Officer is not assigned. Chief Dinosaur says, "During the incident, I don't have enough people to work the fire. There is no one left over to be an Accountability Officer." What the chief fails to realize is that accountability is not something that is done after everything else is covered. If Chief Dinosaur had a water supply problem, he would assign people to take care of it. Firefighter safety and accountability is our number one priority.

Sector Officers are accountable for the companies in their sectors. They must maintain a list of the companies assigned to them, and monitor their activities. Too many Sector Officers don't know what companies are in their sectors because they are never asked to account for them. When accountability reports are held every ten minutes, it forces Sector Officers to be accountable and company officers to keep their crews together. **This is very important**. Sector officer accountability is essential to the success of any fireground accountability system. One progressive fire department has a good accountability system because it takes accountability seriously. They pass out vests and do accountability on every call. They learned as they went along, and made adjustments to the system. "Everyone joked about how goofy they were and how they overreacted," said a firefighter in an adjacent department. "Nobody laughs at them now, because their accountability really works because they use it all the time and they work at improving it. Because they use it on the little things to practice for the big things, they are way ahead of the game."

Sector Officers account for the companies in their sector.

Some departments position an accountability firefighter at the entry points to the fire building from the beginning of the incident. Most fire departments lack the personnel to be able to do this, but positioning accountability officers at the entrances works well when enough personnel become available. As each crew enters the fire building, the Accountability Officer takes their small board, and records the air cylinder pressure and time of entry. If they do not exit when expected, the Accountability Officer checks their welfare. If no contact is made, the IC sends in the rapid intervention team to locate the missing members.

A good accountability system shows the locations of the sectors and companies inside the building. Command must make sure that Sector Officers keep the IC or the Accountability Officer informed of the location of companies inside the building at all incidents. This information is critical if inside crew members become missing.

It is usually easy to record where companies are sent, but firegrounds are dynamic, and it can be very difficult to keep track of a company that goes from Attack to Rehab, to Exposures over time. When there is an Accountability Officer, the Sector Officers notify this officer when companies are moving between sectors. When there is no Accountability Officer, the Sector Officers notify each other and Command when companies are moving between sectors. For example: *Attack to Rehab and Command. Engine 3 moving into Rehab.*

The Safety Officer should not be assigned to handle accountability. If the incident is becoming big enough for Command to delegate accountability, then fireground safety is a full time job. Some departments send a pre-trained command post company to help with both safety and accountability at working fires. This extra company forms two teams, one helps organize the accountability boards, the other monitors scene safety.

The best way to make your system work properly is to make sure that is it fully supported by the brass, use the system regularly and convince the troops that the system saves lives. Making accountability systems work properly is not easy, but it certainly is important.

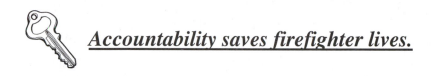 *__Accountability saves firefighter lives.__*

SAVING OUR OWN

Freelancing

Freelancing is a problem in most fire departments. Investigations of fireground operations typically reveal firefighters who were doing dumb things often independently of both their companies and Command. Nobody knew where they were or what they were doing, and they were in great peril. The main reason they lived was because they were lucky. The answer to the freelancing problem is to maintain strict company integrity and accountability, and to have consequences for those who don't play by the rules.

 *__Freelancing kills.__*

Fires Today Are More Dangerous

Fire loads are twice what they used to be. In the past, furnishings were mostly cotton and wood. Today, building contents are polyurethane and plastic. Not surprisingly, studies have shown that fires burn much hotter and build up much faster than in the past. State of the art protective clothing allows firefighters to penetrate much deeper into fire buildings; and many departments have fewer, and less experienced firefighters responding. Add lightweight construction and energy efficient buildings, and you have a recipe for disaster. Fires have changed, but many firefighters have not. The result is that more firefighters are being enveloped in flame than ever before. Tactics from the past can get you hurt. Firefighters need to recognize the following:

- Companies must stay together

- Operating without the protection of hose lines is an increasingly risky operation

- Ventilation is critical, especially in structures with energy efficient double-paned windows

- A very dangerous situation exists when ventilation is incomplete, or when the fire is not being knocked down quickly

- It is very important to have at least two ways out

Why Firefighters Run Out Of Air

Firefighters must be able to recognize situations where the exit time can exceed the time that the warning bell provides.

Most of our fires are in houses and small buildings where firefighters can usually safely exit the building after their SCBA low air bell starts ringing. Firefighters typically wait until the SCBA bell rings before exiting. **WRONG!** The bell is dumb; it just tells you that the air in the bottle is getting low. If you are in a commercial basement, or operating in a maze of office cubicles, waiting for the bell may be a fatal mistake. Your exit time can easily exceed your remaining air time.

I remember being on air at commercial basement fires, closely watching our air gauges with flashlights to make sure that we had enough time to be able to get out. Firefighters must be able to recognize situations where the exit time exceeds the time that the warning bell provides. Streetwise firefighters keep a close eye on the cylinder pressure gauges, and exit (as a team) well before the warning bell rings.

Safety At Problem Fires

We all have fires where operations turn sour in unexpected ways. This causes us to become frustrated, and that can lead to poor decisions. This is a good time for the IC to have another officer at the command post to share ideas.

Here are some examples of the kinds of things that can occur when conditions on the fireground are not going according to plan and frustration and impatience set in:

- Firefighters separate from their companies for independent action or for attachment to another company

DON'T WORK ALONE

- Companies freelance and operate opposing streams at other companies

142

AVOID OPPOSING HOSE LINES

- Firefighters do not leave the structure when ordered out

- Firefighters operate unauthorized hose lines from the outside and endanger inside forces

- Firefighters fail to close doors to protect themselves

In these situations, the companies or individuals involved are not only risking their own lives, but also the lives of their fellow firefighters who may have to rescue them if they get in trouble. When things break bad, keep a cool head, communicate, stick to the SOPs, and don't do anything stupid.

Firefighters In Trouble

In one incident, firefighters in trouble fled from a hot, smoky, hallway into an apartment for refuge, but left the door open behind them. At another fire, the reverse happened. Firefighters fled from a raging fire in an apartment into the hallway, leaving the apartment door open behind them. In both situations, they increased their own risks by failing to close the door for self-protection. This mistake increased the heat and smoke, complicating their escape route.

I was a sector chief operating on the fire floor at a serious office building fire when it happened. I received a radio call from Truck 4, which surprised me because they were not assigned to my sector. Their message gave me chills. They said: *We are trapped in heavy smoke while searching, and we can't find our way out.* I was stunned. I had prided myself on being prepared, but I had never heard a message like this one.

Although we had many companies on the fire floor, it was important that suppression and ventilation activities continue, for the survival of the lost firefighters as well as the rescuers. Stopping these vital activities for a search could easily result in even more trapped firefighters.

Strict control was needed because everybody wanted to help, but they could have made things worse by a well-meaning cavalry charge to find their lost comrades. I informed available companies in the area of the problem and the trapped firefighters were found unharmed a short time later.

This situation shows why it is so important to have a rapid intervention team and an adequate reserve in staging. When firefighters are down, there must be an immediate and predictable reaction. The rapid intervention team is sent in. A rescue sector is created, led by a chief officer, and appropriate backup and advanced life support medical personnel are assigned. The rescue sector should use pre-arranged checklists showing firefighter rescue techniques and procedures that can be used in firefighter rescue.

Bad events can multiply, so make sure you have adequate assistance on the scene. Request an additional alarm for fresh crews because your forces are depleted by the firefighters in trouble and those diverted to search for them. Another standby RIT team needs to replace the one in action. **Develop and practice lost firefighter SOPs.**

Mayday

With air horns screaming evacuation, firefighters began tumbling out of the building. One of the firefighters couldn't find his officer. The rapidly changing and deteriorating fire conditions slowed down Command, so the accountability check had not yet been started. At first, the firefighter thought that the Lieutenant had got out without being noticed. After looking around and not finding him, he became worried and began to approach other officers with his concerns. Their reaction was the same as his initial reaction: "He must be changing a bottle or something." Finally, after the officers couldn't find the Lieutenant, and because the firefighter's concern had changed to panic, they began to react. They reported to the command post that they couldn't find the Lieutenant. The IC reacted the same way: "He's probably outside somewhere." And they started checking around. When he couldn't be found, *then* they started a search and rescue effort. It was now more than ten minutes from the time that the firefighter first noticed the Lieutenant missing. The Lieutenant died in the fire building. Delay can obviously be the difference between life and death.

Following this fire, the fire department initiated a procedure for a Mayday call for firefighters missing or in trouble. Firefighters had to be retrained to react immediately with a Mayday call if they or other firefighters may be in trouble. The fire fighting business has a tradition of rugged individualism. Can do, never quit. We have to change our attitudes so that if you call a Mayday and the firefighter turns out to be okay, nobody says "Ha ha, what boob called the Mayday?" We need to be thinking that we are in a risky business and are better safe than sorry.

Rapid Intervention Teams

The main purpose of this team is to rescue firefighters in trouble.

The RIT is relatively new in the fire service. The main purpose of this team is to rescue injured, lost, or trapped firefighters. Staging areas are not close enough for the immediate rescue of firefighters.

It is important that the RIT is actually used as a rapid intervention team. In one department, the third due engine company is assigned as the RIT. Previously, they used this company for fire fighting. Guess what the third due actually does? When the fire is minor, they are the RIT. When there is a working fire, they do engine work as they always have. This is a paper RIT, and is worse than none at all. If they were used as the RIT at a worker, the fireground would be less safe because they are short a company. This is a ludicrous and dangerous situation.

Your SOP should provide for a rapid intervention team during all interior fire fighting operations. Some fire departments automatically dispatch an additional company as the RIT to all structure fires to get them to the scene quickly. They realize that the first twenty minutes on the fireground are usually the most dangerous. Other departments send the RIT as part of a working fire dispatch or whenever inside attack lines are being used.

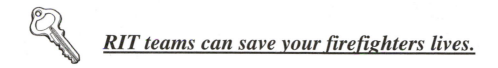 ***RIT teams can save your firefighters lives.***

Rapid Intervention Reality

The RIT should be top-notch people trained in firefighter rescue, not just a team of whoever happens to show up. If there is a choice, truck companies are better suited as a RIT than engines, because trucks normally perform search and rescue duties. This team is usually located near the command post with appropriate equipment, ready for immediate action. The RIT officer must keep the team up to speed on the progress of the operation. When operating in large buildings multiple RITs should be established.

After losing a firefighter in a supermarket fire, the Phoenix Fire Department (PFD) ran a series of RIT tests in vacant commercial buildings and found that the time needed to locate and remove a fallen firefighter averaging around 20 minutes, taking a crew of 12 to remove one firefighter. Rapid intervention in large commercial buildings is often not rapid, and the RIT teams themselves often got into trouble. As a result of the supermarket fire and the follow-up drills the PFD is changing its approach to rapid intervention crews as follows:

- Each first alarm will have a RIT engine and a RIT medical unit

- Working first alarms will have two RIT engines, one RIT Truck company, one RIT medical unit & one RIT battalion chief

- All additional alarms, will receive an additional two engines, one RIT truck company and one RIT Medical unit

One team is sent in to locate the firefighters in trouble. Once located, a second, more heavily staffed RIT team enters to remove the firefighter in trouble.

BUILDING CONSTRUCTION AND COLLAPSE

I was the Lieutenant with Engine 18 as we screamed down Pennsylvania Avenue, first due on the second alarm. There was a column of smoke a mile high, so we didn't need the address. Traffic was light on that Saturday morning, and since the motorists could see the fire, we didn't have much problem getting the right of way.

On arrival, we were told to use our deck pipe. The driver asked me where to place the apparatus. I told him "Just make sure that we don't get close to the walls." We were able to position out of the collapse zone and still hit the fire. Five minutes later, a wall collapsed. We heard no warning. The wall came down with a deafening roar and the ground shook like an earthquake. It was truly a scary moment, and firefighters were injured.

Why weren't we close up like some companies were? I wish I could say that it was my superior knowledge of building construction that kept us out of harm's way. It wasn't. We had a lot of fire in a large, old vacant building, so it was just common sense not to get too close.

The basics are simple. Buildings involved in fire can and will collapse. Most collapses are partial collapses involving floors, ceilings, roofs, walls or cornices. Once a collapse has occurred, expect a second collapse.

The buildings most likely to collapse are:

- Buildings with wide span roofs like supermarkets, churches, restaurants, and car dealerships

WIDE SPAN ROOFS ARE PRONE TO COLLAPSE

- Buildings where the fire is located in the wood or steel trusses used to support floors or roofs

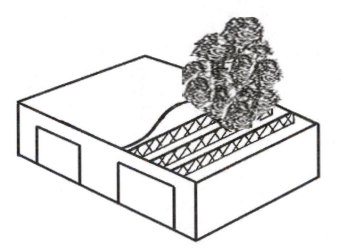

EXPECT COLLAPSE WHEN TRUSS ROOFS ARE INVOLVED IN FIRE

- Buildings having parapet walls extending above the roof

- Buildings heavily involved in fire

WHEN BUILDINGS ARE HEAVILY INVOLVED IN FIRE EXPECT COLLAPSE
Courtesy of Fairfax County Fire and Rescue, Fairfax, Virginia

- Buildings with fire on more than one floor

- Buildings subjected to extraordinarily high fuel loads such as an airplane or gasoline tanker crashing into the building, or buildings containing flammable hazardous materials.

- Buildings of lightweight construction

- Buildings where an interior attack is not making progress

- Buildings that are under construction, demolition, or renovation

BUILDINGS UNDER CONSTRUCTION ARE VULNERABLE TO COLLAPSE

- Buildings that have been renovated through the years

- Buildings with heavy floor loads, such as printing businesses

Know the buildings in your area, and beware of fire situations where collapse is likely. If there appears to be a small fire in the attic or cockloft of any commercial building, and it can't be

147

quickly knocked down with the initial attack line, back out. When you are inside you often can't see the trusses because the ceiling is in the way. A thermal imaging camera can show fire in the attic that is not evident. If you pull ceilings in a commercial building and observe serious fire involvement in the attic or cockloft, report it and get out. When the fire seriously involves the attic or cockloft of a commercial building, do not go under or on the roof. Don't play around in any of these high risk situations. **When in doubt, get out.**

Signs Of Collapse

If you're looking for signs of collapse, you probably shouldn't have firefighters in the building.

Many collapses occur without obvious signs of collapse. We tend to focus on looking for the signs of potential collapse, such as cracks, bulges, creaking sounds, sagging, or smoke and water coming from mortar joints. If you're thinking about looking for these signs of collapse, you probably shouldn't have firefighters in the building.

At a fire in a building that had been converted from residential to offices, company officers noticed a crack in the walls. A collapse zone was set up, and some experienced chief officers and structural engineers ascended in a tower bucket for a closer look. They looked it over and agreed that collapse was unlikely. As soon as they returned to ground level, the wall collapsed, filling the collapse zone. **Assume any doubtful wall will collapse.**

Forget The Twenty-Minute Rule

The truth is you cannot predict collapse time.

The twenty-minute rule said that if we were fighting a fire for twenty minutes, we should consider the collapse potential of the structure. There are several problems with this rule. An obvious question is: when does the twenty minutes start? Does it start from the time you arrive or from the time the fire started (which may have been twenty minutes before the fire department was called?)

Another problem with this rule is that it treats all types of structures and construction features the same. Lightweight trusses may collapse in less than five minutes under fire conditions. The truth is you cannot predict collapse time. I recall a fire in downtown Washington D.C., when the entire side wall of a building collapsed abruptly without warning, about thirty seconds after the arrival of the first company.

 <u>Recognize collapse situations and do not enter those areas.</u>

Collapse Zones

When setting up master streams, anticipate structural collapse and set up collapse zones. Use yellow tape or cones to mark the collapse zone, and enforce it with a Safety Officer. Strict control to prohibit entry is necessary because many firefighters don't seem to think that the collapse zone applies to them. The collapse zone is the full height of a wall, plus one third, to allow space for flying bricks and spill over. Example: A 15-foot high wall can collapse its full height, plus flying bricks and debris out to a distance of up to about 20 feet from the base of the wall. Don't forget to consider a collapse zone for the aerial tower bucket when it is used for elevated master streams The bucket must be positioned where it could not be struck by a falling wall. **The Moral: Don't even think about going into a collapse zone.**

For more information about building construction and collapse refer to Frank Brannigan's classic, <u>Building Construction for The Fire Service,</u> 3rd Edition, published by the National Fire Protection Association.

MAINTAIN A DISTANCE OF THE BUILDINGS' HEIGHT, PLUS ONE THIRD

PROTECTIVE GEAR

Check your gear before each use, wear all your gear and use your safety equipment. Your SOPs should mandate consistently using protective equipment during drills and calls. Practice wearing all your equipment and it becomes automatic. Automatic use allows maximum protection at every fire.

Pass Alarms Are Useless Unless...

Many PASS devices do not work until you turn them on. Don't laugh, because this is a deadly serious problem. The U.S. Fire Administration reports that about three-fourths of the firefighters who died in the past five years after being caught or trapped during interior fires had not turned on their PASS devices.

The reason for the problem is complacency, one of the main problems in the fire service. What we do at our routine calls is what we tend to do at the big incidents. Most of our calls are minor and don't seem to require the use of a PASS alarm. When the big fire happens, or if the routine fire suddenly turns sour, the PASS alarms are not armed, because it is not a part of our regular routine. In one city, an explosion occurred while firefighters were conducting a seemingly routine investigation. If the firefighters had activated their PASS alarms, it would have been much easier to locate them in the debris. Since firefighters are dying with them turned off, it is obvious that we have many firefighters who routinely work without the alarms being turned on. How do we get people to turn them on? It's simple. Make a big deal out of it. Educate, train, monitor, and enforce. This is a deadly important safety issue.

Fortunately technology is catching up, and PASS devices are now being incorporated into new SCBA, so that the pass device is activated without a deliberate action by the wearer, typically when you turn on the air supply.

We Can't Find You

Don't wait until a firefighter is down to organize your firefighter rescue system.

Finding a lost or fallen firefighter after the PASS alarm sounds may not be easy. It's like trying to tell which direction a siren is coming from. It sounds simple, but you can be easily fooled. PASS alarms echo in buildings making locating the alarm difficult. Many PASS alarms sound like building alarms. In addition, different brands of PASS alarms have different pitches, so you need to get used to the PASS alarm sounds. In addition, the flashing light on newer PASS devices may not be easily seen, depending upon the circumstances.

At one house fire some of the troops were running out of air. They exited the building and some left their SCBA on the front lawn. The PASS alarms that were not turned off began to scream. At the same time firefighters inside the house were in trouble, but their PASS alarms could not be heard over the outside noise. It is important to turn off PASS alarms when they are no longer needed.

Conducting lost firefighter drills regularly and using SOPs are the best ways to be prepared for firefighter rescue. Thinking about this problem in advance and developing a plan is much better than freewheeling. PASS devices are a new technology for the fire service, and it often takes a while before new procedures are incorporated into our manuals and routine. Don't wait until a firefighter is down to organize your firefighter rescue system. This is a top priority, life and death issue.

All Firefighters Should Wear Hoods

Nomex hoods must be an important part of the protective clothing system. They protect areas not covered by facepieces, collars, or earflaps. Question: If you knew that on your next fire call that you were going to be caught in a flashover, would you wear a hood? Of course you would. You must wear one routinely, because your next call could be a flashover. Some fire departments do not issue hoods, and some even have rules that prohibit firefighters from wearing unofficial equipment. This means that firefighters protected by hoods bought at their own expense can be subject to disciplinary action. Brilliant.

FIREFIGHTERS MUST WEAR HOODS

Let Your Light Shine

Many firefighters are kept in the dark. They work without hand lights or rely on the lights of others to be able to see. This is a dangerous practice that must not be allowed. A personal light that is hazardous atmosphere approved must be issued to each member. All new apparatus should be ordered with rechargeable hand lights for all firefighters.

ALL FIREFIGHTERS NEED HAND LIGHTS

Smart Firefighters Don't Breathe Poison

There seems to be an unwritten rule for firefighters to take off their facepieces as soon as possible. It used to be that old time firefighters breathed poisonous gases during the fire, for lack of SCBA. Now firefighters breathe poisons after the fire. In past years, the smoke from fires was from wood and cotton. Now the smoke comes from plastics and vinyl. Every house fire today produces hazardous materials smoke. As soon as the first kamikaze firefighter takes off his facepiece, the pressure is on everybody else. Nobody wants to be the only one to have the facepiece on. Did you ever notice that when the facepieces are off, the tanks remain on the backs? That's because the firefighters aren't sure that they can hack it without the masks. Having air on your back and breathing poisons doesn't make sense, but it happens every day. Firefighters can't do this without a green light from their officers and chiefs. It is a command responsibility to look out for the safety of the troops. Too often in the fire service, we monitor and enforce mickey mouse things, but we allow our people to breathe poison because we always have done it that way.

USE MASKS, DON'T BREATH POISONS

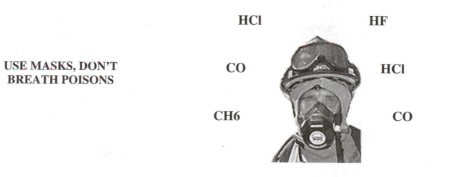

HCl   HF

CO   HCl

CH6   CO

Testing The Air

Air testing is not Star Wars. There are a number of relatively simple and inexpensive devices available that measure toxic gases in the atmosphere. Chief Dinosaur says "We can't afford gadgets." The cost of one firefighter's visit to an emergency room will probably exceed the cost of the air monitor. These monitors show concentrations of fire gases, such as carbon monoxide and polyvinyl chloride (PVC), given off by burning plastics. Whenever SCBA has been used or toxic gases are involved, readings should be taken. The firefighter doing the testing should be using SCBA to avoid being a canary.

TESTING THE ATMOSPHERE

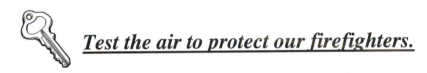

Test the air to protect our firefighters.

What You Can't See Can Hurt You

We owe it to our firefighters to do what we can to reduce the hazards they face.

Chief Dinosaur says, "Any dummy can see when the smoke is gone." True, but you can't see the fire gases. One afternoon we had a copying machine fire in an office building. After knockdown, but before the smoke was out, air monitoring readings were taken. The readings showed low levels of PVC gas present. Just before we were ready to leave, with no smoke visible, we took another reading and much to our surprise the second readings were *higher* than the original readings. We talked to haz mat people afterwards. They said that what happened was not that unusual. I don't understand it, but I don't have to. I don't guess. I just test the air. All fire department SOPs should require firefighters to keep on their SCBA until the atmosphere has been verified as safe. Unfortunately, this is rarely done. Why? Because it is relatively new, and conflicts with our "but we've always done it this way attitude."

A Smart Self Contained Breathing Apparatus (SCBA) Policy

Fire departments should have clear written policies for the use of SCBA. A simple basic SCBA policy follows:

- An SCBA should be promptly used for smoke, electrical fires, and hazardous materials incidents including unknown odors

- Air monitoring readings should be quickly taken after knockdown. Following the air tests, Command notifies all companies to either continue using SCBA or to discontinue mask use because the atmosphere is safe

152

- For the safety and protection of all members, SCBA and facepieces will not be removed until the "all clear" signal has been broadcast

- When overhauling involves insulation in ceilings, walls and around pipes, assume that asbestos is involved and use SCBA

Post Fire Safety

Overhaul begins when the safety crew reports the area to be safe.

One situation that makes the hairs stand up on the back of my neck is when everybody stampedes back into a building following defensive operations. The place has already been subject to major fire damage, and massive weights of water, and the stability of the building is doubtful. After the master streams have shut down, take some time to let things settle down and conduct a thorough safety check before considering re-entry.

The safety check should be a standard practice that follows all working fires. After knockdown and control, get everyone out of the building and conduct a safety check before overhaul and salvage begin. Select one crew of qualified people to conduct the safety check with the Safety Officer. This crew looks at the stability of walls, floors, and roofs. They check for holes in floors, missing stairs, electrical or gas hazards, and the presence of asbestos. Dangerous areas should be taped off and proper lighting provided.

CONDUCTING A POST FIRE SAFETY CHECK

The post fire safety check should also look for the possibility of hazardous materials, including clandestine drug labs. Checking avoids ugly surprises like the one we had following a house fire when we were unknowingly shoveling hazardous materials during overhaul. We were unaware that the fire was in a crack house, and because of the presence of drugs, we were in the middle of a haz mat incident, thinking that we were at a fire. The situation went unrecognized because we had assumed that house fires and hazardous material incidents were different situations. Don't assume that structure fires and haz mat incidents are unrelated.

The post incident hazardous materials check is proactive, getting ahead of the curve to avoid ugly surprises. Too often at these incidents, we start to think about hazardous materials during overhaul when firefighters start to complain of headaches and skin rashes; then it's "uh-oh" time, and we have to scramble to handle the problem. Overhaul begins only when the safety crew reports the area to be safe.

Post fire safety checks prevent injuries and save lives.

It's easy to forget safety after the fire is out. An example is a fire that burned in a large vacant building all night. In the morning, a chief ordered the companies to start pulling the plywood off the first floor windows. They were banging away on walls five stones high that had been exposed to heavy fire and tons of water. A Sector Officer, seeing the insanity of the operation, asked the chief, "What are we trying to do here chief?" hoping that the chief would realize the absurdity of the situation. The chief paused for a long time, and then said that he was ventilating. As soon as the chief went around the corner, the companies stopped.

Dangerous situations can arise when overhauling a residence as well. When there are spot fires burning in cornices and eaves, firefighters will sometimes risk their lives operating in the collapse zone trying to extinguish some leftover fire in a burned out building. **Don't allow overhaul to take priority over firefighter safety.**

It is not written in stone that fire departments must overhaul at all fires. Some fires are not safe to overhaul, for example when there are walls and floors that are not stable. Avoid the temptation to overhaul because that's what you usually do. In doubtful situations, consider what you have to gain against what you can lose before beginning any overhaul. Ask yourself, "Is it really important that we send our people inside?" Consider what would happen if you don't overhaul.

I recall one major fire where the building was in a dangerous condition. An exterior wall had already fallen during the fire. A senior officer leaving the fireground told me, "Be careful when you overhaul." I was careful all right, I didn't let anybody near the place. We used a ladder pipe until the wreckers leveled the building. Sometimes ladder pipes and towers can be used for a safe and very effective hydraulic overhaul. When it is too dangerous to overhaul, a fire watch is usually maintained until the building is made safe or demolished. When overhaul is necessary after major fires, consider waiting for daylight before beginning. Use fresh troops to overhaul if possible. There is more about this in the De-Escalation chapter. **The moral is: Don't get hurt after the fire's out.**

Apparatus Accidents

With the air horn blaring and siren screaming, the heavy rescue squad entered the intersection. The squad lieutenant never saw it coming. The tractor trailer truck hit them broadside, spinning them down the side street. The unbelted squad driver was thrown into the street. He missed by inches the rear tires of both the rescue squad and the tractor trailer. Miraculously, he fell between the vehicles and suffered only minor injuries.

The best way to avoid injuries from vehicle accidents is by using seat belts. Belted-in firefighters walked away from their apparatus that was rolled over several times after a collision with a cement truck. "In my department" said one firefighter, "our crew cabs allow firefighters to ride facing backward, belted in with their masks at their backs for easy placement." But, he complained, "Most firefighters ride in the forward-facing jump seats, without seat belts on, so they can look out the windshield. Then, when mask-up time comes, they have to stand up in the cab to put their masks on. If they are ever in a bad accident, they will bounce around the crew cabs like rubber balls, if they're not thrown outside."

You would think that because firefighters respond to so many vehicular accident calls they would become more safety conscious. You would also think that the officers and chiefs would strictly enforce seat belt rules, but too often they just do not. It will probably be business as usual in these lax departments until somebody gets killed. Then using seat belts will be a big deal—too late.

We have too many apparatus accidents in the fire service, most are caused by driver error. Too many departments allow the guy who annually wrecks his own car to go behind the wheel of fire apparatus. Some departments allow firefighters to drive with little or no formal training. One department teaches recruits to drive on the city streets. After one trainee wiped out three parked cars, the investigating police officers couldn't believe how the fire department teaches new drivers. We tend to focus more on the hardware, like the color of the apparatus, instead of the software in the drivers' heads. Crazed apparatus drivers can cause us problems even when they get to the scene without an accident. The response is often so scary that personnel are upset and rattled even before they get to the incident. These troops are behind the eight ball before they even start.

The best way to avoid accidents is to put the right people in the driver's seat after training them in an emergency vehicle operator's class. Sounds simple. It is simple, and it will reduce the rate of accidents considerably, perhaps dramatically.

Roadside Accidents

Every vehicular accident scene is another accident waiting to happen.

FIRE CAPTAIN KILLED IN ACCIDENT, screamed the newspaper headline. The captain was operating at the scene of an accident caused by ice on the roadway. An oncoming car hit the same patch of ice that caused the original accident and slid into the rescuers, killing the captain.

It is a sad but familiar story, because this double accident is not uncommon. If you stop and think about it, the reasons are obvious. Traffic accidents often have contributing factors, such as wet roads, ice, snow, and blind curves. The conditions on the scene are ripe for another accident. Unfortunately, firefighters are often nonchalant about the dangers that await along the road.

How can we avoid these tragedies? We must realize that every vehicular accident scene is another accident waiting to happen. We must operate as if the next car approaching the scene has a drunk driver who is speeding. It just might. Rule one in arson investigation is to protect the evidence. Rule one at accident scenes is to protect the scene.

ONCOMING TRAFFIC →

PROTECT THE ACCIDENT SCENE

Fire trucks make great protective barriers. Establish a safe working area by parking apparatus at an angle, blocking the scene with your apparatus so your apparatus becomes your shield. One fire department, after losing a firefighter at an accident scene, now sends a ladder truck to all accidents and vehicle fires along with an engine company. The truck is positioned in front of where the engine crew is operating. A major function of the ladder truck is to be a big red wall that is all lit up. It works. Some departments have equipped their apparatus with permanently mounted large arrows that light up and flash as warning beacons. Many enclosed crew cabs give firefighters the option of exiting the rig on either side. Firefighters should always get out on the side away from the traffic after pausing to look around for any imminent hazard, and the apparatus should be positioned so that the officer's side is away from traffic if possible.

Have you ever sent firefighters to place roadside warning flares, and they were back in thirty seconds. They went a few feet down the road to set flares. Oncoming traffic needs a lot of advance notice and reaction time. Use many flares and start setting them way, way back from the scene. Set out the first flares or cones at least 350 to 1,000 feet depending upon road conditions.

Sometimes roads must be shut down for scene safety. The police usually don't want to block roads. They want to keep traffic moving. Moving traffic is a big priority for them, but our top priority is firefighter and scene safety. It may help to have departmental SOPs for closing roadways and to meet with the police officials to explain our policy, so we don't have to argue with the street officers at the scene.

The typical firefighter wanting to stop traffic at the scene or in front of the fire station will step out into the path of oncoming traffic with his hand up. **WRONG!** If the driver is drugged, drunk, or not paying attention, you're dead. You don't stop trains by standing on the track waving a hanky, and you don't stop traffic with your body. Stand out of the way when you stop traffic. Do not stand in the path of oncoming traffic. As one firefighter once told me, "If they can't see this big piece of apparatus lit up like the fourth of July, they sure as hell can't see me." **The moral is: If you get hit by traffic, make sure they have to come through a fire truck to do it.**

SUMMARY

Safety needs to be taken seriously. Every fire department must evaluate their safety practices to narrow the gap between what they say and what they actually do.

Fire departments should know how their safety records compare to those of similar fire Departments.

Injury and accident statistics should be used to develop plans to reduce injuries.

Commitment to safety equals success. Get the safety message across by publicizing and rewarding good safety records.

Most vehicle accidents in the fire service are caused by driver error. The best way to avoid accidents is to put the right people in the driver's seat and train them well.

Before taking action at an emergency scene, we need to consider what we stand to gain against what we can lose. We need to recognize that after the occupants are safe, our primary consideration is the safety of the firefighters.

It is very important that officers keep their companies together. Each crew must enter together, work together, and leave together.

Your accountability system must be simple, accurate, and used on every call.

At routine incidents, the IC is normally responsible for accountability.

At working incidents, the IC usually designates an Accountability Officer to handle accountability.

Firefighters must be able recognize situations when the exit time can exceed the time that the low air warning bell provides.

Organize your firefighter rescue system and conduct regular lost firefighter drills.

A rapid intervention team should be standing by at working incidents for firefighter rescues.

Learn to recognize potential collapse situations.

Buildings involved in fire can and will collapse.

If you are thinking about looking for the signs of collapse, you probably shouldn't have firefighters in the building.

Fire departments must have clear written policies for the use of SCBA.

We should not allow our people to breathe poisons.

Mask facepieces must be kept on until air testing shows an "all clear."

Before overhaul begins, select one crew of qualified people to conduct a post fire safety check. They look at the building stability and safety.

12

PRE-PLANS

"Avert the danger that has not yet come." - Yoga Sutra, 2.16

Pre-plans ensure that fire departments know as much as possible about a building's construction, occupancy, hazards, and protection systems before an incident occurs. Pre-plans save time, avoid confusion, reduce the hazards to firefighters and can make the difference between success and failure. Water supply should also be pre-planned and mapped out to help establish and maintain an adequate water supply.

Operating Without A Pre-Plan

When you are operating with poor information, your effectiveness is limited. One afternoon at a warehouse fire two firefighters fell into a pit built into the floor. Shortly after this, two more firefighters fell into a separate pit. Fortunately none of them were injured. The IC and the firefighters were unaware that there were holes built into the floor. Pre-plans help make occupancies less like land mine fields.

I recall one blaze where I thought the fire was in one building with exposures on both sides. After the fire, I found out that it was all one building without exposures. I also thought that this was a three-story building, but it was actually four stories. Commanding a fire is difficult enough if you know what you are up against. When you are operating in the dark with poor information, your effectiveness is limited. This means that the fireground is much less efficient, and more hazardous for the operating forces.

Making Pre-Plans Work

Pre-plans should be aimed at target hazards that have the potential for major life or property losses. Examples of target hazards are hospitals, strip malls, hotels, large apartment buildings and structures with unprotected steel truss roofs. One and two-story residential structures seldom need to be pre-planned, since they are usually handled by the department's standard operating procedures.

Focus pre-plans on what is vital and avoid filling pre-plans with unimportant information. Remember that the plan will be used under difficult conditions, and that the incident commander doesn't have time to read much material. A one-page written description, with a diagram of the building and its surroundings that can be reviewed and absorbed in less than one minute usually works best. The key to pre-plans is to have the right information and be able to use it under emergency conditions.

Pre-Plan Basics

It is important that pre-plans:

- Be completed by people who are sold on the pre-plan idea and who will likely be there if a fire occurs

- List the type of occupancy and the type of construction

- List the building materials and components of the structure

- Include a sketch of the basic floor plan and a diagram of the building showing the building's access, as well as the shape, height and dimensions of the building; and include locations of any exposures

- Use standard pre-fire planning symbols to reduce the amount of writing and make the diagram easier to read

- Clearly show the levels in buildings that have entrances at different levels, to avoid confusion during fire fighting operations

- Highlight and outline the major hazards and problems, such as those involving the presence of hazardous materials, lack of standpipes, interconnected exposures or conditions that may affect firefighter safety

- Include basic water supply information, such as which streets have large mains or where a drafting position is located

- Be easy to retrieve on the fireground: the state of the art approach is an on board computer, but a well-indexed loose leaf notebook carried by the fire companies also works

- Be reviewed routinely and updated on a regular basis, during walk through inspections conducted by the fire companies

- Be tested by conducting on-site drills

Pre-Plan Problems

The main reasons why pre-plans are not used include the following:

- The plan is often in the middle of a pre-plan book that is neither handy nor indexed

- Plans are often too wordy and complicated. The command post is a very busy place during the early stages of an incident, and the IC doesn't have time to read very much about the structure

- In the heat of battle, pre-plans can be forgotten, since we seldom use them. We don't have many fires in target hazards, so they are often not part of our routine

- If the IC fails to delegate using ICS, he may become too busy to refer to a pre-plan. The pre-plan is typically needed when we have a working fire in a major building, which is a very difficult and busy time in the life of the IC

160

- Pre-plans are often out of date. If the IC knows or suspects that the pre-plan is not current, it may not be used for fear that no information is better than bad information.

Pre-plans are important, do them right.

A sample pre-plan follows:

ADDRESS: 123 Oak Street

OCCUPANCY: Auto Body Shop

SIZE: 30' by 40'   1 story

MAIN HAZARDS:

Roof collapse, open floor pit quadrant C and hazardous materials quadrant B.

CONSTRUCTION:

Concrete block, with unprotected steel bar joist roof supports. Concrete slab floor.

HAZARDOUS MATERIALS:

Acetylene bottles, paints, and solvents in quadrant B.

BASEMENT: None

FIRE PROTECTION SYSTEMS: None

WATER SUPPLY: Hydrant, 800 GPM, SW corner 3rd and Oak Streets.

UTILITIES: Electric meters on rear wall.

EXPOSURES: Exposure B, 1 story similar construction, distance 25 feet.

This pre-plan is bare bones basic, but that is what is needed. Once the plan is written, see that everybody involved, including mutual aid companies, has a copy available for immediate use.

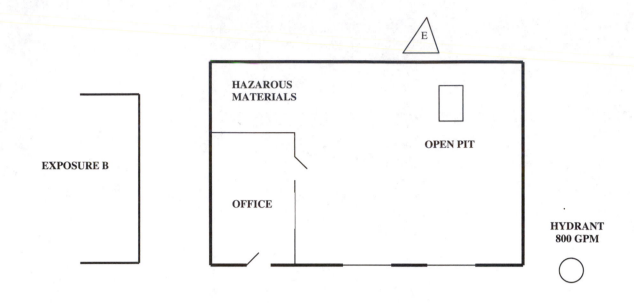

ACE AUTO BODY 123 OAK ST

SUMMARY

It is important that pre-plans have the right information and be easy to retrieve and use on the fireground.

Pre-plans should include the basic building information, and highlight major hazards that may affect firefighter safety. Pre-plans should also include water supply information.

Pre-plans should be aimed at target hazards that have the potential for major life or property losses.

Pre-plans must be updated on a regular basis.

13

COMMAND

"Those who can command themselves, command others." - William Hazlitt

It was a bitterly cold night. We responded to a store fire in a mixed commercial and residential district. The fire was in the basement of a corner store that had two floors of storage above. This fire happened a long time ago when I was a newly promoted chief. This was before ICS, accountability, command charts, fixed command posts and the rest. We were in the dark ages of fire fighting, sometimes referred to as "the good old days."

During the first few minutes of the fire, I managed to get hit with a hose line as I ran around the fireground and was instantly turned into a walking icicle. Things weren't looking very promising, so I called for a second alarm. There was a small apartment building attached to the fire building, so I assigned an engine company to protect this exposure. Soon my boss, the deputy chief, arrived, along with the second alarm response. We fought a long hard battle, and finally we won the battle, or so we thought. The deputy and I were down in the basement thanking the troops for their efforts. Suddenly we heard screaming that the fire was coming through the roof.

We headed for the street and looked up at the fire which was now worse than when we first got there. The deputy reached for the radio.

> *Deputy Chief to Dispatch.*

> Go ahead Deputy.

> *Send a third alarm.*

A stunned silence followed. We had been there over an hour and a half and everybody including us was expecting the fire to be winding down, but the fire had other ideas.

> Repeat your message.

> *Send us a third alarm.*

> Are you are asking for a third alarm?

> *Correct, send a third alarm.*

I quickly checked with the engine in the exposure. They weren't there. They said, "It looked okay in there a while ago, so we left it." I had neither checked back with them, nor assigned companies to the floors above the fire. I told them to get back into the exposure. Within a minute, they replied "We have fire in the exposure." Quickly, I assigned another company to check the next attached exposure down the line. That company soon reported heavy smoke and possibly fire in that building. Next to that exposure was an attached ten-story apartment building that could barely be seen through the billowing smoke. The occupants were bailing out carrying suitcases.

The deputy turned to me and said, "If the fire gets into that big apartment building, I'm going to retire in the morning." It didn't, he didn't, and the fire was controlled again—or so we thought. We were now about three hours into this fire from hell, and the troops were bone weary and frozen. We called for fresh companies who normally proceed to the fireground obeying traffic regulations. Before the relief companies got to the scene, the fire started to get away again. We didn't have enough people or energy to stop it. I had to call Dispatch back to have them tell the relief companies to respond to the scene; we needed them in a hurry.

Eventually the fire went out, but we were worn out and demoralized. Needless to say, this wasn't our finest hour. In our defense, this fire was the exception for us; most of our firegrounds went reasonably well. Before you get too critical about our lousy, yahoo operation, remember that the problems that plagued us continue to be problems for many fire departments to this day.

This fire was a learning experience. As the years passed, I became a more experienced chief, and we starting using effective command systems. Now the same fire would be handled much more efficiently. We would have created an exposure sector, and with the additional focus on exposures, we would have realized that the fire was moving up through the partitions and into the exposures. We would have had an adequate reserve in the staging area. We would have rotated the crews into rehab much faster. The fire would have been confined to the original fire building and probably would have been contained in the basement. Some of what I learned about command in the subsequent years follows.

What The IC Does

The principal jobs of the IC are to:

- Size-up

- Set up a stationary command post, using command charts and check-off systems

- Consider the gains and the risks

- Use SOPs

- Determine the plan

- Establish sectors to implement the plan

- Coordinate the sectors

- Assign a Safety Officer and a rapid intervention team when necessary

- Protect exposures on all sides, inside and out

- When the outcome is in doubt, call for help quickly and establish a staging area

- Set up logistics to handle incident needs

- Designate aides to assist Command as needed

- Transfer command as appropriate

If the troops know that the IC has a plan, and is firmly in control of the situation, they can go about their fireground tasks with a greater sense of security.

COMMAND PHILOSOPHY

Over a beer, the new battalion chief began telling me about a fire he had commanded that went south big time. There were water problems, size-up problems, ventilation problems, communication problems and a mayday for trapped firefighters. He told me that he didn't think it was possible to have all those problems happen so quickly at one fire. He also said that he had had a previous fire where there were missing firefighters who had miraculously survived. He had only been a chief for two years. Knowing that I had been a fireground commander for fifteen years, he wanted to know if I had had any similar experiences with firefighters in trouble. I told him only once as a sector chief, but never as the incident commander.

I told him my philosophy. My job is to keep these things from happening. I do it by making sure that the troops are doing what they are supposed to do on every call; I don't accept sloppy. I tell my company officers, up front, exactly what is expected of them, so they don't have to guess. I make sure that firegrounds work properly, and I watch everything like a hawk. If something isn't right on the fireground, I'll have a little talk with the officer. I never raise my voice or get angry (well, hardly ever), but I am persistent and I am the boss, so they all get the message sooner or later. Guess what? I don't have many problems on the fireground because I constantly work to keep the little problems from becoming big problems.

SIZE-UP

At most fires there are only a few key points that are really important. The trick is to focus on those that are critical to the incident.

Proper size-up sets the tone for an incident by identifying the problems. This procedure in turn helps to determine the initial strategy. Life safety of both firefighters and civilians is always the first priority. The location of the fire is also important. For example, basement fires have a greater potential for loss of life and property than top floor fires. If the fire is in the partitions, extinguishment will likely be more difficult and require increased personnel. Consider the building's occupancy: for example, a fire in an auto body shop may involve hazardous materials. Whether the building is vacant or occupied will affect your strategy.

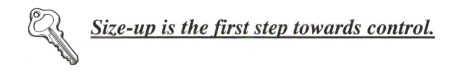 *Size-up is the first step towards control.*

A FIRST FLOOR FIRE HAS MORE POTENTIAL THAN A TOP FLOOR FIRE

When it's feasible, arriving chief officers can take a 360 degree ride (or walk) around the building to size-up the situation before assuming command. The Attack Sector Officer should report inside conditions to Command. The outside and inside size-ups can be compared to help access the situation. The IC has many factors to consider, and the list is a mile long. At most fires however, there are only a few key points that are really important. The trick is to focus on those that are critical to the incident. If the IC starts down the list of all the possible factors, the place will probably burn down before he gets to the last consideration.

Offense Or Defense?

The IC must conduct a risk assessment of the incident before beginning operations. Since most fires are fought from the inside there is a tendency to do what we usually do. It is extremely dangerous to have troops inside a building when it should be a defensive operation. Missing the signs of a defensive operation can be fatal. These signs usually include the following:

- The structure cannot be saved

- The fire is on more than one floor

- There is little to gain, such as abandoned building fires

- The fire is in the trusses, and cannot be knocked down quickly

- There is doubt about the stability of the building

A Screwed Up Size-Up

Consider the following size-up considerations at a recent fire:

- The fire was in a one-story commercial occupancy with a trussed roof

- The fire occurred in the middle of the night

- Forcible entry was difficult, requiring almost ten minutes to gain access

166

- The roof team reported heavy smoke and fire conditions from the roof and fire in the ceiling area

- When engine companies were finally inside, they reported high heat, zero visibility and no fire

The really important size-up considerations were:

- The fire involved the truss roof of a commercial occupancy

- The life safety of the firefighters

The best way to ensure firefighter safety was to anticipate collapse and implement a defensive operation, but that's not what happened. Command stayed with the interior attack for ten minutes, and ordered the evacuation only after the troops began to self-evacuate because of rapidly deteriorating conditions and a roof collapse. All the firefighters did not get out on time, and many of those who did get out were lucky.

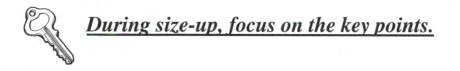

During size-up, focus on the key points.

COME UP WITH A PLAN

The IC must quickly decide the goals and then develop a plan to accomplish them. The IC who delegates through sectoring will avoid being overwhelmed and will have some "thinking" room. The IC must consider, "What is the situation, and what are we trying to accomplish?" That may sound goofy, because the answer to the questions may appear obvious. However, if you go up to the typical IC at a fully involved abandoned house fire with no exposures and ask what is the plan is, the IC is likely to look at you like you're crazy and tell you that they are trying to put the fire out. **Time out!** The building is already lost, and there is nothing to gain. If the IC had considered "what am I trying to accomplish?" The answer would be obvious. There is nothing to be gained at this fire; the main goal is to operate safely.

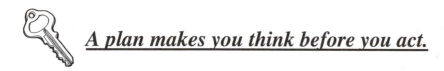

A plan makes you think before you act.

A Planning System

Some fire departments formalize the planning process by using a system to help determine the proper strategy and tactics. By using a system, the IC doesn't have to make all the decisions "out of thin air." This can be a big help, especially for new officers and chiefs. The simple system shown below (which can be part of the command chart) can be used at any structure fire. The risk factors are circled for the fully involved abandoned house fire discussed above.

RISK FACTORS

FIRE STAGE	LOW	MEDIUM	**(LATE)**
SUCCESSFUL RESCUE ODDS	**(LOW)**	MEDIUM	HIGH
FIREFIGHTER RISKS	LOW	MEDIUM	**(HIGH)**
STRUCTURE STABILITY RISKS	LOW	MEDIUM	**(HIGH)**

These risk factors circled above are then used to determine the basic fireground plan as circled below.

FIREGROUND PLAN

RESCUE	INTERIOR ATTACK	**(SAFETY FIRST)**	PROTECT EXPOSURES	**(EXTERIOR DEFENSIVE)**

All officers on scene or en route must have this type of planning system in their mindset. You shouldn't have to wait for a chief to go through the risk/plan factors. Too many officers are wired for, "Damn the torpedoes, full speed ahead." There is a time for this, but not routinely. These officers will charge every time unless a chief tells them otherwise, then they *might* change modes. **Fire officers need to think and plan, not just react.**

Using a fireground plan will result in consistency and improved firefighter safety. Review all plans periodically using both observation and feedback from the sectors. Consider the time factor when making plans because the time between making a decision and implementing it may take longer than you think, causing your plan to be too late to do any good.

Plans should be simple with clear and realistic goals. Don't try anything too complicated. Avoid trying new things during the incident; stick with what is familiar to the troops. Officers should be aware of the plan and assigned broad objectives, not given step by step instructions. When major decisions are made, they should be announced so that the personnel will know what is happening.

Plans are one thing, but having the resources to implement these plans is another. The key is to call for help early and have plenty of companies in staging. A full staging area can be the difference between success or failure, because deteriorating conditions usually require much more people and equipment.

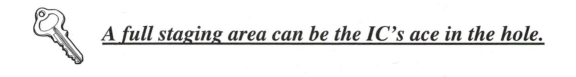

A full staging area can be the IC's ace in the hole.

Let The SOP Carry You

Good SOPs make the life of the IC much easier because during normal fireground operations, routine work is done automatically. For example, a backup line is used for safety and firepower because the SOP requires it. It happens without orders from the IC. When SOPs cover routine situations, directions from the IC are needed only when the circumstances are not routine. This is called management by exception, and it allows the IC time to think and plan ahead. The IC must stay sharp to recognize situations where normal SOP operations do not apply or won't be effective. **The moral is: Management by exception allows the IC time to think and plan.**

On routine fires using good SOPs, there is little for the IC to do except see that things are going as expected. Some chiefs have problems with having little to do, and they want to micro-manage every incident. That works until the incident begins to escalate. When fire conditions are getting worse and a lot is happening, the IC who must tell everybody what to do will quickly go down in flames (pun intended.)

Backup Plans

A backup plan assumes the worst and anticipates that the original strategy could fail. The backup plan avoids putting all your eggs in the same basket. It is hard to develop a backup plan; most IC's have all they can do to come up with one plan. Usually, alternate plans are needed for incidents that are escalating, but that is the busiest time for the IC. By using ICS, SOPs, an Operations Officer and an aide, the IC can take some time to look at the big picture and develop a backup plan. For example, at the typical structure fire, alternative plans may include considering the possibility of more victims than anticipated, or anticipating changing from offense to defense.

The IC Should Be Flexible

The incident commander should be flexible and open to recommendations. Don't let your ego get in the way of good advice. Be reasonable and approachable. The IC needs to listen to feedback from the sectors. You have many good people in your fire department and many heads are better than one. Obviously you can't run an incident by consensus, but the idea is to make things work as well as possible, not to show who is the boss or play turf games. All IC's have been in the position of having a company officer suggest something so obvious that you felt stupid for not having thought of it yourself.

A chief in an area of attached townhouses had a fire in a store with an apartment above it on the second floor. The fire in the store was smoky and difficult to find. The IC asked for a second alarm. There was a crying need for more ventilation, lights and relief in the store.

What did the IC do with the second alarm companies? He assigned them all to exposures which were not severely exposed. Why? Because that is what he normally did on the townhouse fires. He was used to calling for help and assigning the second alarm to protect the exposures to the fire, not to the fire itself. The effective IC is flexible and tailors the response to the specific situation. Plans should be flexible and updated during the operation.

If interior companies aren't making progress, try different approaches; have others explore an attack from another position or location.

COMMAND POSTS

The command post also provides a place to retain relative sanity.

The stationary command post allows the IC to be a true fireground manager. The command post has status charts and check-off systems because you can't command properly if you don't know the sectors, companies and status of the incident. The command post provides a strong apparatus radio to receive and transmit messages, so you can practically eliminate problems such as lost messages and having to repeat messages three times to be understood.

THE COMMAND POST DURING A MAJOR OPERATION

Quoting a chief who had operated from a fixed command post: "If I had left the command post I probably would have been sucked into the operation, smoked up, drowned and been in a lousy position to command—particularly if the fire had escalated." Another IC who had set up a command post on the sidewalk in front of an apartment building fire stated. "I should have stayed in my vehicle across the street, because I became overwhelmed by the people exiting the building, and I was fast becoming an Information Officer."

"The command vehicle is my office. It's the best way to run the show." (Battalion Chief Edward A. Rinker.) You can do a better job in a warm, dry and quiet environment, than you can standing in the street in the pouring rain, with your running coat half on, trying to answer companies with your portable radio, with air horns from incoming companies blowing in your ear. Staying in the vehicle has many advantages. It's quieter. Sirens, revving up pumps, hollering and yelling, power saws and breaking glass do not make for the best environment in which to concentrate to make cool and calm decisions.

Incident Commanders must operate from a fixed command post.

From a properly positioned command post you can usually see whether or not the fire is being hit. You can see if the fire is extending up or sideways, or threatening exposures. You are in a better position to see the need for help and to position incoming companies. Some departments have stories of pump operators calling for help, because the IC missed the obvious by being off someplace running around.

I recall one fire where I had a good view of the fire building from inside the command vehicle, but I was unable to see any of the smoke or fire. Command still worked well because my radio network gave me an accurate picture. The only other possible positions were in the street (where I still wouldn't have been able to see anything, and my communications would have been poor) or in a smoky stairwell or on the no visibility fire floor (where the only thing I might be able to command was myself.) **The moral is: You can think better and bigger using a command post.**

The IC Cannot Run And Think

Many fire departments have operated for years with a roving IC, and some still do. It is very difficult for them to understand that there is a better way to operate. The IC's who like roving command are those who have always done it that way, and don't know any better. My experience has been that IC's who have worked with both roving and stationary commands strongly favor stationary command.

Using mobile command (running around with a radio), the IC can see more, but can't accomplish as much. The roving position causes the IC to waste time going from one place to another, and, possibly, to become sidetracked. It is very difficult to run around and think. It is even more difficult to run around and establish priorities and a plan. Sectoring allows the IC to remain at the command post because the Sector Officers become the eyes and ears of the IC.

One department experienced a fire where things went terribly wrong. Command was located in a poor position and could see very little. Communications were so poor that they couldn't even hear much of what was happening. Command was clueless. Naturally, word of this problem fire spread quickly through the department. The reaction of some chiefs was, "I'm not going to let this happen to me. I'm going back to running around the fireground with my portable radio." They totally missed the point that the problem was not with the command system, but with the commanders who didn't use the system properly

The IC and The Door Hinges

It is easy for a roving chief to get involved in a company operation, a lesson that I learned the hard way. While roaming at a working fire, I happened upon a crew that was having difficulty forcing a door. I got involved in the entry operation. Think about it. I was the IC at a working structure fire, and I was concentrating on a set of door hinges! The job of the IC is to look at the big picture, not at door hinges. The problem with my being preoccupied meant that there was no command post. No one knew where I was except the people forcing the door. Even worse, I didn't know where my companies were. Since I was depending upon a portable radio in the middle of a forcible entry operation, my communications were poor. I could not hear the radio very well, nor could I be easily understood when I sent messages.

My "hinge" fire occurred years ago, and I learned my lesson. However, the problem still goes on in the fire service. I read recently where an assistant chief of a large department was directing a forcible entry operation at a high-rise fire.

It can be hard not to rush in and become a Super Captain. One argument that you hear from roving chiefs is that an IC's years of experience are wasted at a fixed command post. The thinking seems to be that the wise old IC may see things that junior officers may miss. Yet, most chief officers believe that they have generally good and experienced company officers. Company officers don't need the IC to be peeking over their shoulders. They can handle the fire, but they can't manage it; only the IC can do that. Most firegrounds have more problems with weak command than with ineffective company officers. **If you want to lead a company, be a company officer.**

Limit Access To The Command Post

Operating inside a vehicle limits the number of people that you have to deal with, which makes the IC's job easier. Access to the command post should be limited. On major incidents, the area should be sealed off, and a command post "guard" should be posted to allow only authorized personnel access to the command post.

One Command Post Is Enough

It is very important that all fireground requests come from the command post. When radio channels get too busy and Sector Officers are unable to contact the command post, they may get frustrated and bypass the command post. This results in messages such as: *"Water supply to Dispatch, send me another engine company."* Likewise, medical personnel sometimes start an independent operation on the fireground. This can result in messages such as: *Medic 1 to Dispatch, send another ambulance.* Avoid radio system gridlock by using ICS; avoid multiple command posts by using radio discipline.

Problems can also arise when a senior officer arrives and does not take command. Often the senior officer who isn't in command starts to get involved in command post operations. Before you know it, the troops start talking to the non-IC. You soon have a monster with two heads. Senior officers should stay home, keep quiet or take over.

COMMAND CHARTS AND CHECK-OFF SYSTEMS

The most practical chart is lean and mean, with just the basics, and nothing fancy. Complicated charts don't work well during fast moving and complex operations. There are many commercial charts, or you can make your own. The three basic types of charts are: paper, plastic, and metal. With paper charts, you can print them to fit your needs and have copies made. Plastic can be written on with a marker and re-used for other incidents. Magnetic charts use "chips" for each company. The chips are already labeled with company numbers and sector names which saves time, and they can be easily moved around without erasing. This can be important at fast moving operations with many companies. Handwritten charts can quickly become a jumbled mess under these circumstances.

Use Command Charts To Keep Track Of Companies

It is very important for Command to keep track of sectors and companies for firefighter safety and proper incident management. To use the command chart, the IC **must** know what companies are responding. The IC who charts every incident develops good management habits and will do well when an incident escalates. The IC who waits for the big one before using the command chart faces mental meltdown.

Incidents should be charted from the beginning of the operation. Too many departments don't keep track during the early stages. The thinking seems to be that it is not necessary to track a few companies. They think that tracking becomes important only if the fire escalates. The problem is that it's a poor time to get organized when conditions are getting worse. Recording events as they happen is easier than playing catch up and trying to reconstruct events later.

KEEP TRACK OF ALL THE COMPANIES

Often, a great deal of activity takes place while the chief officer is responding. Unless there is an aide driving (which is unlikely), the chief often winds up making notes on a yellow pad lying on the seat. Some chiefs don't record anything in the mistaken belief that getting to the scene quickly is more important than being on top of what is happening. The chief officer often arrives in the middle of the initial attack with nothing on the command chart and a bunch of scribbled notes on a pad with no time to transcribe them on the command chart.

The chief typically goes on using the yellow sheet. At a house fire involving a room, the yellow pad will probably work. But using a yellow pad most of the time gets the IC in the habit of running incidents by using a yellow piece of paper, not a command chart. If the fire escalates, the yellow pad is inadequate, and command starts to unravel. The answer is to record information on the command chart as it occurs while responding. It may be necessary for the chief to pull over while responding to start filling out the command chart, but it is time well spent to be fully prepared upon arrival. Remember, the chief organizes, and the companies hustle.

During interior operations, keep track of where companies are operating inside the building, not the location of their apparatus. Your concern is where the crews are, not where the pumper is sitting. Sector Officers must provide regular status reports on their location and progress. If there is any doubt about where companies are, the IC or the Accountability Officer should confirm their location. During master stream operations, the control chart should show where the apparatus is positioned, and the location of crews who are not with their apparatus.

 Command officers should use charts and check-off systems on every alarm.

Check-Off Systems

It is very important to use a check-off system to ensure that everything is covered in a methodical and timely manner.

Airline pilots are required to use a checklist every time they land. The copilot reads to the captain each item on the list. The captain then performs the task, which is then checked off. This is done on every landing; it doesn't matter how experienced the pilots are. The reason for this rule is that some plane crashes are caused by pilots forgetting to put the wheels down or some other necessary step.

We need similar checklists in the fire service to ensure that everything is covered in a methodical and timely manner. The fireground is too important to leave to memory, and it is impossible to remember everything without some type of system or checklist. The check-off is usually part of the command chart. The check-off system has proved itself at firegrounds in many fire departments. Since it works so well at our bread and butter fire operations, it stands to reason that it will work for other types of incidents, such as those that involve hazardous materials, water rescue, extrication, and so on. We normally don't run many of these types of incidents, and lack of experience with a situation makes it all the more important to use a check-off system.

Using A Check-Off System

The check-off assures that the important points in an operation are covered.

The check-off shows which tasks are completed, and which remain to be done. If an important point is overlooked, a glance at the chart will reveal it. This ensures that all the important points in the operation are consistently and promptly covered. Actions reported by the sectors are checked off on the chart, along with the time, if possible. For example, if early in a fire the electricity is reported as shut off, you check it off. Later, before overhauling, if you are asked the power status, you have a record that the power is off. Without a check-off system, you will not be sure, and we can't guess about the electricity status. If any of the items on the check-off list do not apply, they are crossed out. At the end of every incident, all the items on the chart will be either checked off or crossed off. This system avoids the fate of one IC who, following a fatal fire, was cited for failing to realize that he had never received a search status report.

Your system should be set up so that the Sector Officers give the IC all needed reports. Your troops must be programmed to give you status reports on every call. Your job is tough enough, and you shouldn't have to ask for everything that you need. If you don't get the required reports, never let it slide. Follow up and ask for the report. Later you can remind the responsible officer about the importance of the reports.

What Happens When You Don't Use A Check-Off System

Without a check-off system, the IC is not on top of what is happening, and important things can fall between the cracks. You can tell there's a problem when company officers or Dispatch starts making suggestions for things that are needed but haven't been considered or requested. Typical suggestions are, "Will you need relief companies?" or "Could you use foam?" or "Do you need an air truck or a lighting truck?" Using a check-off system avoids overlooking important considerations because the system prompts the IC to stay on top of the needs of the incident. A sample check-off follows:

STRUCTURE FIRE
CHECK-OFF CHART

ATTACK LINE (   )
BACKUP LINE (   )
BASEMENT/ BELOW (   )   REAR (   )
VENT (   )   ROOF STATUS (   )
RISK/GAIN (   )   PLAN (   )

SEARCH   1 (   ) 2 (   )
RAPID INTERVENTION (   )   RESERVE (   )
DIAGRAM (   )   AIDE (   )

ACCOUNTABILITY CHECKS (   ) (   ) (   ) (   ) (   ) (   )

SECTORS ESTABLISHED

EXPOSURES A (   ) B (   ) C (   ) D (   ) ABOVE (   ) INTERIOR (   )
SAFETY (   )   LOGISTICS (   )   MEDICAL (   )
WATER SUPPLY (   )

TOWER (   ) MASTER STREAMS (   )

TIME/ RELIEF/ REHAB (   )
MASK SUPPLY (   )
SALVAGE BELOW (   )

CALL FIRE INVESTIGATOR (   )
       ELECTRIC (   ) GAS COMPANIES (   )

SAFETY CHECK (   ) TEST ATMOSPHERE (   )
     CEILINGS/ATTIC (   )

Using Aides — No More Lone Ranger

The IC cannot manage a big incident alone.

Every IC has had the feeling. Things happen so fast and furious that soon you are drowning in a sea of overload; even when using ICS and sectoring. Is the building occupied? Where is the fire? Where are the exposures? It is very important for the IC to have adequate information. A surprising number of IC's really aren't sure exactly what the situation is, and commanding without adequate information is difficult and dangerous. The IC usually needs help with the radio and charts to be able to get a handle on the situation. At all large incidents, every IC needs an aide or aides to help manage the incident.

During fireground simulation exercises at the National Fire Academy, the IC is always assigned an aide. The reason is simple: the IC cannot manage a big incident alone. It is important that personnel who serve as aides be familiar with the ICS and command chart operations.

THE AIDE AT A WORKER

At a fire in a commercial building, a firefighter showed up at the command post and volunteered to be the aide and assist the chief. Fireground communications were poor and the aide was reporting valuable fireground information to the chief. Unbelievably after awhile, the chief told the aide that he didn't need him anymore. At this point, fire conditions were deteriorating, and the chief wasn't doing too well with someone helping. The fireground quickly became your worst nightmare. Why did the chief turn away help that he desperately needed? Most IC's operate without an aide at most fires, so an IC may tend to do what is usually done, and become the Lone Ranger. The IC who fails to use an aide or delegate at a major incident will roll over and go under with the first wave.

What The Aide Does

The aide usually starts by going around the involved building or area to report smoke and fire conditions, exposures and safety concerns. The aide also reports to the IC the dimensions, height, occupancy and configuration of the structure. A three-story building in the front can become two stories toward the rear. Buildings can be L-shaped or T-shaped or have unusually long dimensions, and it may be that none of these characteristics are visible from the command post. The aide should sketch a diagram of complex situations. The aide works for the IC and should not get sucked into hose or ladder work. Sample reports from the aide follow:

This is a two-story detached house, 25 x 40, heavy smoke sides B and C, both floors— fire may be in the basement.

Building is 40 x 60, two stories, looks abandoned, heavy smoke, top floor quadrant C (right, rear)—appears to be a room, attic looks okay, no exposures.

Building is three stories 100 x 100. Fire showing third floor quadrant B (left, rear), looks like an apartment off. The exposure is a similar three-story apartment building 25 feet away on side B (left side.)

After completing the size-up, the aide should work with the IC at the command post, helping to handle radio traffic and the command chart. Fireground aides also relate to the IC pertinent information from pre-plans or personal knowledge about the building.

Diagrams are needed at the command post during working incidents. The diagram comes from a pre-plan or an aide and provides a "snapshot" of the incident for the IC. When the situation is confusing, one picture is worth a thousand words.

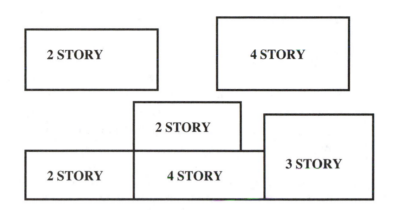

TRY EXPLAINING THIS ON THE RADIO

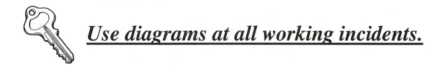 *Use diagrams at all working incidents.*

Who Is The Aide?

This is a good question that doesn't have a quick and easy answer. Ideally, an aide is regularly assigned to assist the chief, but, for most of us, that is not going to happen. Some volunteer departments may have members who are not able to do nozzle and hook work. They could be used as aides. Sometimes, just calling for help at a working incident can get enough people to allow the IC to be able to designate an aide. Some departments use staff personnel called from the office to lend a hand. Other departments use medical personnel, who are standing by in case they are needed. Sometimes a company is used: one member becomes an aide and the others are assigned to staging or safety. Some departments send an additional company to working fires to assist at the command post. There are many ways to provide a fireground aide, and each department needs to develop a system to provide an aide at working incidents.

COMMUNICATIONS

The IC must communicate with the Sector Officers to stay on top of what is happening. This is basic, but it doesn't always happen. I have a fireground audiotape in which the IC never checked with the Attack Sector on the progress of fire control. In another case, the IC never heard from the first due engine for over an hour and made no attempt to contact them. The engine company's crew of three had become trapped and died. Your system must make sure that all the key players check in at regular intervals. **The moral is that communications are vital.**

Reacting To Important Messages

The IC should never simply acknowledge things that require action.

When important messages are received such as: *the stairs are in a dangerous condition,* all sectors must be notified of the hazard, and each sector must acknowledge it. Just telling the troops about a hazard isn't enough. Some may either forget it after a while, or it will go in one ear and out the other. Command should react to reduce the danger by having the stairway made safer or blocked off. When the safety problem can't be corrected, clearly identify the hazard area with cones or tape, and enforce it with a Safety Officer.

Not Reacting To Critical Messages

Good communications require listening and reacting to messages. If a company reports: *We have victims rescued from the fire lying on the back lawn,* sometimes Command acknowledges and then goes on to other messages. **Time Out!** This message is important and requires immediate action. An appropriate response would be for Command to establish a medical sector, and to make sure that there are enough resources to handle what has been reported. Command should also anticipate what could happen. For example, if the searches are not completed, there may be other victims still inside, so the IC should maintain an adequate reserve.

If a sector requests relief from the IC, then a relief crew should come. Too often, requests and suggestions are ignored. The IC should never simply acknowledge things that require action. A major reason this happens is because the IC is overwhelmed by radio traffic.

I listened to a radio tape in which an officer on the upper floor of a high-rise building reported to Command: *I am lost and I can't find my way out of here.* Command acknowledges the message, but there are no messages about rescue operations. I listened to the tape with an officer who was at the incident. My initial reaction was of disbelief. Then I thought that attempts had been made that were not recorded on the tape. His disgusted reply was, "No, unfortunately the radio tape tells the whole story; no rescue attempt was ever made." When I asked how this could happen, the reply was, "We don't get many fires. Our command is generally weak, and this is what happens when you have a roving command post." The officer eventually found his way out, no thanks to Command.

You can avoid this problem by improving training and communications, and by using rapid intervention teams and trapped firefighter SOPs. Operating from a fixed command post strengthens command, as does having an aide to share the work load. The command chart will show which sectors and companies are closest to the firefighters in trouble. All of these factors will help the IC to react to critical messages quickly and predictably.

RECOGNIZE EXTREMELY DANGEROUS SITUATIONS

Certain situations on the fireground scream danger. The IC must be alert to these extremely dangerous situations and make the right decisions to make sure that nobody gets hurt. Examples of these dangerous situations that kill and injure firefighters follow:

One Way In And One Way Out Fires

When firefighters are doing inside fire fighting, make every effort to provide at least two ways out. Avoid one way in and one way out situations. Create a second way out by opening up doors and windows and raising ladders to the upper floors. If companies are operating in a boarded-up building with only one entrance, pull the boards off the windows and doors for safety and ventilation. Providing adequate ventilation will assist interior crews in getting out quickly if necessary.

BEWARE OF ONE WAY OUT FIRES

Attic fires usually have only one small exit that may be hard to find. Use attack lines in attics only as a last resort. Make sure that the attic is well ventilated, and that a backup line and RIT team are used. Place a light at the attic exit as a guide.

When a hose line is advanced up a ladder and into a fire building, firefighters must retreat down the ladder if conditions deteriorate. Avoid this situation unless you are performing a rescue.

Look Out For Companies Above The Fire

When companies are operating above the fire, the dangers are obvious. The IC should make sure that these crews have adequate ventilation, at least two exits (ladder the building) and a charged hose line. The IC must maintain radio communications with sectors above the fire and monitor them very closely.

LOOK OUT FOR COMPANIES ABOVE THE FIRE

179

Beware Of Fire Fighting In A Maze

Search lines should be used by crews operating in these areas.

Fighting fire in a maze can be a firefighter's nightmare. Mazes are often found in areas sub-divided into cubicles for business use and in industrial occupancies. Sometimes the maze is the result of the high stacking of stock, with the added collapse hazard. The area doesn't necessarily have to be large to present problems. Sometimes 'pack rat" residences create mazes. Fire fighting crews operating in mazes can have difficulty finding their way out, and it is easy for firefighters to get lost or separated. I recall one office building maze-type fire where it was difficult to find our way out after extinguishment and ventilation was completed.

The IC must maintain strict control at these types of fires. Lighting and ventilation are crucial. Make every effort to ventilate quickly to reduce smoke and heat conditions. Lighting is important, and lights should be placed at the exits as beacons. Search lines should be used by the crews operating in these areas, and accountability of personnel must be closely monitored.

MAZES ARE A FIREFIGHTERS NIGHTMARE

<u>*Firefighters should use search lines and position bright lights as exit beacons when operating in large or maze-type buildings.*</u>

Hazards Of Very Large Structures

Crews experienced with house fires who find themselves operating in a smoky Super Wal-Mart can easily get into serious trouble.

Crews operating in very large buildings are in danger of becoming lost and unable to find their way out. The possibility of running out of air is another danger. If the square footage of the building is much larger than what the crews are used to operating in, then it's a large building.

Crews experienced with house fires who find themselves operating in a smoky Super Wal-Mart can easily get into serious trouble. The IC should handle these fires like the maze fires above, because they are very similar. Crews should never leave the fire attack line or search rope when operating in large structures.

**BEWARE OF FIRES IN VERY
LARGE STRUCTURES**

Hazards Of Non Residential Structures

Operating at commercial building fires puts you are at much greater risk than operating at residential fires. Fires in manufacturing facilities, storage facilities, stores and offices, places of public assembly (including churches), abandoned buildings, and buildings under construction or demolition have very high firefighter death rates. **Operating at one of these buildings makes you three to five times more likely to die than operating in a residential structure.** Common sense tells us that it is more dangerous to operate in these structures, and the statistics bear this out. Since most of our fires are residential, and comparatively few occur in commercial occupancies, we often fail to recognize the increased danger in commercial structures.

 **Be extra careful when fire fighting in non residential structures.**

NON RESIDENTIAL OCCUPANCIES ARE VERY DANGEROUS
Courtesy of Fairfax County Fire and Rescue,
Fairfax Virginia

Multi-Level Entrances

Companies really don't know where they are.

Years ago I was on a hose line at a commercial building fire. My engine company had entered the building by a side entrance and then down some steps. Smoke was heavy as we advanced our line in what we thought was the basement. We could see the fire and were moving in on it when we realized at the last second that the fire was coming up through the floor. We almost fell into it. We found to our horror that we were on the first floor, not in the basement where we thought we were. Sloping ground caused the building's entrances to be at different levels.

In many parts of the country, it is common for dwellings to have two stories in the front, but three stories in the rear with a walk-out basement. When you go in the main entrance of one large apartment building in Washington D.C., you are actually on the third floor. Entrances at different levels can be very confusing. Some buildings have a mezzanine level located between floors. This also causes confusion. Reports about fire conditions and company locations in these buildings will often be wrong, because the companies don't know where they are. This creates a very dangerous situation, and the resulting confusion has caused firefighter injuries and fatalities.

Knowledge of the building and a good pre-plan are the best defenses. The IC can suspect that multi-level entrances are a problem when receiving location messages that don't seem to make sense. When confronted with a multilevel entrance problem, it is very important for the IC to notify all sectors and companies to make sure that they are aware of the problem.

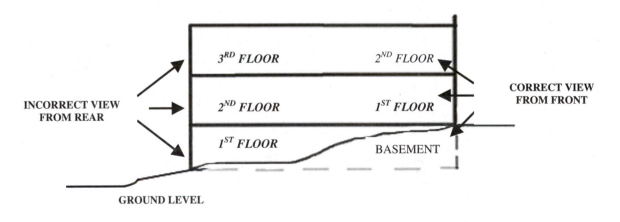

BEWARE OF MULTI-LEVEL ENTRANCES

Not Your Normal Fire

In most of the country, the majority of structure fires are in single family houses. Some areas have mostly row (or town) houses, some mostly apartment buildings. The result is that most fire companies normally fight the same types of fires in the same types of buildings. These "normal for you" fires can usually be expected to go reasonably well because of prior experience. Problems can arise when the row house fire companies have a fire in a split level house, or when the split level company has a fire in an apartment building. Chief and company officers need to be extra careful when they are faced with fires in unfamiliar types of structures.

Following a tragic fire in which firefighters became lost and died, the IC said: "We did what we always do: we attacked, vented and searched. We have done it a thousand times before and it always worked. I don't know what we could have done differently." The problem was that almost all of their prior experiences were in residential structures. The fire was in a huge multi-story building with large open areas. Unfortunately, they used their house fire plan.

BE EXTRA CAREFUL WHEN IT IS NOT YOUR NORMAL FIRE
Courtesy of Fairfax County Fire and Rescue,
Fairfax Virginia

When The Building Is Not Properly Vented

Fire fighting inside poorly ventilated structures is dangerous. The failure to ventilate keeps smoke and heat inside the building. This reduces visibility, increases the dangers of flashover or backdraft, and reduces the chances of survival for both victims and firefighters. Proper ventilation reduces smoke and heat, improves visibility and can save firefighter lives.

When You Can't Knock Down The Fire

If the fire can't be knocked down, or you can't find the fire within a reasonable period, it's time to evaluate the risks versus the gain.

It is dangerous and very frustrating when an interior attack fails to result in a quick knockdown of the fire. Sometimes, the reason is that crews are unable to find the fire. When this is the case, thermal imaging devices can be helpful. Increasing ventilation often helps locate the fire because lessening smoke and heat conditions should make it easier to find the fire.

Sometimes, the reason for failure is that there is too much fire to extinguish quickly. Adding another fire attack line will sometimes solve this problem.

 **At problem fires, increasing ventilation and putting more water on the fire may help.**

When the location of the fire can't be found, Command must consider the possibility that the fire could be below the troops and they could drop into it, or that the fire is overhead and could collapse down onto the firefighters. Another possibility is flashover or rapid fire buildup that could trap firefighters. **Too many firefighters have been injured or killed because they remained too long in fire buildings.**

Don't screw around with fires. If the fire can't be knocked down, or you can't find the fire, it's time to evaluate the risks versus the gain. Often, you are gaining nothing, and the risks are great and increasing by the minute. The fire is destroying the stability of the building, so the longer you are inside, the greater the risk.

EVALUATE THE RISKS WHEN YOU CAN'T KNOCK DOWN THE FIRE

Abandoned Building Fires

Avoid a cavalry charge.

Abandoned buildings can be deathtraps for firefighters. Many firefighters are wired to go full speed ahead. It is difficult to get them to slow down and realize that abandoned buildings aren't worth the risks. Fight these fires from the outside, if possible. If you must go inside, wait until the building is thoroughly opened up and vented. Avoid a cavalry charge by limiting the personnel allowed inside, after calming them down. When entry is made into an abandoned building, it should be slow and deliberate. Easy searches can be carefully made, but avoid "Hail Mary" searches in these buildings. Carefully weigh the potential gain against the safety risks. Guidelines for entry into abandoned buildings should be part of your fire departments SOPs.

BEWARE OF ABANDONED BUILDING FIRES

Firefighters Under Wide Span Roofs

Any fire under or in a wide span roof is a potential killer. Typically these roofs are found on supermarkets, outlet stores, churches, car dealerships, and other structures with wide, open spaces inside. Unless the fire is small enough to be easily and quickly extinguished by a single hose line, back the troops out and go into a defensive operation. To remain is to court disaster.

184

DON'T WORK UNDER A FIRE IN A WIDE SPAN ROOF

When Everything Seems To Be Going Wrong

We all have incidents when nothing seems to go right, and things don't go according to plan. Command and the troops are doubly stressed, sometimes resulting in dumb decisions made out of desperation. Did you ever notice in the movies, just before something bad is about to happen, the music becomes creepy and "spider music" starts to play in the background. This music starts to play at fires when things aren't going well. You can hear it if you listen carefully, the feeling of impending disaster is almost palpable. In these situations, changes are needed. The IC should call together a few experienced officers to discuss the options. The IC can then make the necessary changes to try to get things back on track.

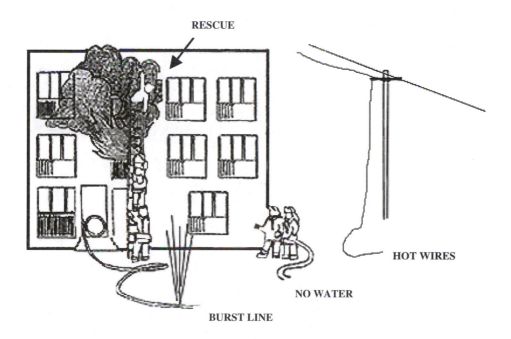

BE EXTRA CAUTIOUS WHEN THINGS ARE GOING WRONG

185

Lost Causes

Above all, look out for the safety of the firefighters.

Unfortunately, in the fire fighting business, we don't win them all. The simple fact is that sometimes we will lose buildings through no fault of our own. We have a can-do philosophy in the fire service that is great in the right situations. We normally take aggressive action to make things better. The problem is that at lost cause fires, aggressiveness can get somebody killed. We must avoid doing dumb things when trying to turn around a losing situation. At these times we need to say—realistically— we are going to lose this attic or building, but let's do it with a little bit of class, without Keystone cops heroics and without injuries. Commanders must communicate and maintain fireground discipline, and, above all, look out for the safety of the firefighters. When all else fails, get everyone out of harm's way, start a good aquatic show and set up your command post at the canteen.

**DON'T GET HURT AT
LOST CAUSES FIRES**
Courtesy of Fairfax County Fire and Rescue,
Fairfax Virginia

Ways To Reduce The Danger

Consider reducing the hazard by avoiding the situation or by taking action to make the situation less dangerous. An example of risk avoidance is deciding that sending a crew up a 35-foot ladder and into a small attic window is too dangerous. This is often a command decision because Lieutenant Kamikaze and the troops may be eager to take these kinds of chances.

Conservative tactics oriented toward protecting the firefighters should be used when confronted with dangerous, unknown or unanticipated situations. At all dangerous situations, a Safety Officer should be used and a rapid intervention team should be standing by for the possible rescue of firefighters.

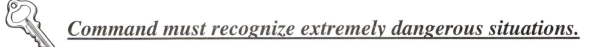

Command must recognize extremely dangerous situations.

CALLING FOR HELP

When you think big, you are prepared for incidents that escalate. You can scale back much more quickly than you can upgrade. The philosophy of the IC should be to have an adequate number of companies to accomplish the task, and to have reserve for unforeseen events. During working incidents, there should always be companies staged in reserve. Having a reserve avoids radio transmissions such as: *Exposures to Command, we need two more engines,* Followed immediately by: *Command to Dispatch, send me two more engines.* Obviously no companies were in staging, and those just called will take time to arrive (maybe too late to do any good.) **The moral is: When the outcome is in doubt, call for help.**

Don't Let This Happen To You

I had no reserve, mostly because the situation had not appeared serious.

Nothing was showing when the first companies arrived at a small apartment building. I was the IC, and the companies reported light smoke in the building. The voices on the radio were not excited; in fact they did not even sound concerned. Soon they said that there was a little fire in the wall on the second floor. Over the next ten minutes, I had committed all my companies to check the floors below and the attached exposure next door. The whole situation was deceptive because there was never much smoke or fire showing.

Then, I had two back-to-back requests for more companies. The fire floor needed another engine and truck company, and the exposure company reported fire in the exposure. I had no reserve, mostly because the situation had not appeared serious. Help was a long way off. This was a learning experience, and I vowed never to get caught short like that again. I don't like learning from my mistakes. I'd rather learn from other people's mistakes. Hopefully, you can learn from mine and be prepared.

Calling For Help Shouldn't Be A Crap Shoot

The most important reason to call for help is for life safety.

Too often, calling for help varies greatly depending on who the IC happens to be. One eastern city usually calls for help quickly, figuring that it is better to have help and not need it than to need help and not have it. Some departments and some chief officers have reputations for being stingy about calling for help. There are many reasons for this, including pride (false or otherwise) and tradition. Some departments seem to have a rule that the last firefighter able to stand up calls for help. Some don't call for help because of politics. One Assistant Chief didn't want to call for third alarms because, in that department it meant notifying the mayor. Some chiefs may not want outsiders coming in to "their" fire. In big cities, some chiefs have been known not to call for help when it is needed because they hate their boss, who will come with the next alarm.

But We Always Handle It Without Help

The wrong philosophy is trying to make do, hoping to make it without help. The idea is to handle the incident properly, not conserve fire companies. Some IC's seem to be excessively concerned that if they use too many companies, it will mean reduced fire protection in their area. One of the recent California brush fires brought a response by 440 engine companies. You read it right, *four hundred and forty fire companies.* Adequate protection was maintained through mutual aid. Are you still worried about fire protection for your area when you have a working house fire? The truth is, that we are all a part of one big fire department, and we have to readjust our thinking to accept this fact.

I know of a six station department that handles their larger incidents by using no more than four stations, if they possibly can. Years ago they often had to go it alone, but now they are in a county-wide dispatch system. They are really part of a large 25 station fire department, but they don't seem to realize it.

I recently talked with an officer from a western state who works in a small career department. I asked him how they manage to handle serious fires, since the area is so remote. He said that they call back off duty personnel. He added that the closest department, which is considerably larger, was about 45 minutes away. The way he said it made it sound as if the help was coming from China, and that, by the time they could arrive, it would be too late. The last time they had called for outside help was fifty years ago! It is safe to say that there have been fires in the last half century in this department when outside help should have been called in, and perhaps it could have affected the outcome. They think that the problem is simply geography, but the problem is more the department's philosophy. They think they are so remote that they must be self-sufficient.

In major cities, it is not unusual to call for fourth or fifth alarms after an hour or more into the incident. Major city departments have help arriving well into the incident. Why do some smaller departments fail to call for help because it may not be there quickly? Perhaps the answer is tradition—a failure to adapt to changing times.

Flow Charts Are Lousy Indicators Of When To Call For Help

Some folks recommend using flow charts as guidelines for calling for help. Flow charts are usually lousy indicators of when help is needed because all they consider is water volume. The flow chart formula may show that you will need 750 GPMs for a building that is 25% involved in fire. This may require two engine companies for extinguishment. If the 25% involvement is in one, easily accessible area, and there are no other considerations (not real world), then two engine companies should be able to extinguish the fire. But, if the 25% of fire involvement is in the basement, in the partitions, or on the thirtieth floor, many more than two engines are necessary.

Another shortcoming of flow charts is they only consider engine companies. Structure fires usually require ventilation, rescue, medical, search, staging, logistics and more. None of these are addressed by flow charts even though they are essential to the fireground operation.

Requirements of the Incident Command System and NFPA 1500 increase our personnel needs. In the past, we often didn't have Safety Officers, logistics, RITs and other positions that are often needed to manage emergency incidents properly. None of these important positions are addressed in flow charts.

In addition, many of the flow chart systems originated during the days when staffing was much greater than it is today. Using the flow chart to determine personnel needs is misleading because the number of engines required does not consider staffing. Fire companies today often have half the numbers of firefighters as in years past. The bottom line is that flow charts can give rough water requirements, but are usually poor guidelines for calling for help.

Help? I Knew I Forgot Something

In the middle of the night a passerby knocked on the firehouse door to report smoke in an area near the fire station. The engine and truck company responded to investigate. I was the chief on duty, and I listened for the radio report of what they found, but it never came. I then listened to the fireground frequency and heard it crackling with their fire activity. I responded and found that they had a room on fire on an upper floor of a large apartment building. They were in way over their heads, and I immediately called for a full response. After the fire, I talked with the two company officers about the importance of calling for help quickly. To my surprise, they were defensive and argued that they didn't need any help because they had the fire knocked down before the rest of the assignment arrived.

They thought that they had done a hell of a job (they had) and were looking for praise, something like, "Wow, you guys did a great job considering you were so shorthanded." Instead, I told them that they'd made a mistake and that they should have called for help. I explained the reasons why it was a bad decision: the life hazard, the time of day and other things I didn't think that I should have to explain to fire officers. They never did really see the light. It seems that some people have to have an operation blow up in their faces before they change their approach. Unfortunately, too many chief officers suffer from the same short-sighted vision.

I once witnessed a downtown fire in a row of stores. The initial assignment was two engines, a truck company and a chief. Heavy smoke was showing in the front, and a company in the rear reported fire coming through a portion of the roof. The IC called Dispatch, and I was sure that he would call for help. **WRONG!** The chief requested the response of a squad. I couldn't believe it. The situation screamed for more help, which seemed obvious to everyone at the fire except the IC. Eventually the fire wound up a third alarm, but it took a long time. The IC was no dope and was experienced, but failed to act. I think that we all have had the experience of listening to a working fire on the radio and just *knowing* that the IC needed help fast; even though help was called for much later if at all.

One night, I responded as the acting deputy chief to a working fire in a furniture store in the far reaches of the city. The area was primarily residential, so they didn't get many store fires. Before I left the station, I had an aerial tower dispatched, along with an additional engine company for its water supply. I thought this was a reasonable backup considering the type of fire and its remote location.

Going over the road I listened to the fireground radio chatter, and it was not encouraging. I heard transmissions such as, " the fire is over here now" and "we need more lines." Things obviously weren't going well. When I was about halfway there, the chief used the engine company that I had sent to supply the aerial tower, assigning them to go inside the building with hose lines. When I heard that, I called for a second alarm. If the chief was using companies he hadn't call for, they were in trouble.

The fire was contained and didn't require any help beyond the second alarm. The next day, the regular deputy whose place I had been taking asked me about the fire. He couldn't believe that I had called for help when I wasn't at the fire and there was another chief on the scene. I explained to him what had happened. He asked, "What did the chief on the scene have to say about what you did?" I told him that the other chief had said, "Thank you."

Why do these things happen? One reason is that these IC's probably thought they were doing the right thing because those before them (from whom they learned) sometimes did the same thing. Another reason is that most chiefs just don't get that many fires. Even when they do have fires they get by without calling for help most of the time. This can lull chief officers into complacency. In addition, a chief who has been coasting through the minor incidents can be overwhelmed by the big incident. Firegrounds get very busy very fast. If ICS is not properly used, radio traffic will balloon out of control, and the IC quickly becomes a radio operator. Under these circumstances, the whole command process bogs down, including calling for help.

Calling For Help Quickly

If you need them now, and haven't already called for them, it's too late.

If you use ICS, you are likely to call for help more quickly because the blank spaces on the command chart tell the IC that a lot more help is needed to fill the positions required. At the downtown fire in the row of stores discussed above, the control chart below shows the need for a lot of help.

"Hell's fire," says Chief Dinosaur, "You don't need no chart to tell you that the damn place is burning down." That may be true but the chart will show what has been done and what remains to be covered. This gives the IC a rough guide to how much help is needed, and the wise IC will add to this to be on the safe side.

ATTACK	VENT/RESCUE	EXPOSURES	MEDICAL	LOGISTICS	SAFETY
E 1	T 1				
E 2					

THIS COMMAND CHART SHOWS THE NEED FOR A LOT OF HELP

Another consideration in calling for help is how long it will take the help to get there. If you are out in the boondocks or the weather is bad, it is better to call for help sooner than later. Firefighters want to go to fires. Get them on the road.

Some fire departments send automatic second alarm assignments when companies arrive to find smoke or fire showing from high-rise buildings or other target hazards. These incidents need lots of help fast, and the automatic second alarm provides the help needed for effective operations.

Alarm Times Can Tell A Story

Whenever you read fire stories in the fire magazines, carefully note the alarm times. The timing of an incident can tell you a lot about the department, the IC and, sometimes, the results. One example was a fire in a commercial building in a downtown business district. The initial alarm was at 00:43 hours. The second alarm was at 01:08, twenty-five minutes into the incident. If it took five minutes for the department to arrive, the battle was fought for about twenty minutes before the IC called for help. Was there an adequate reserve if something went wrong, such as flashover or missing firefighters? Probably not. The fire eventually became a fifth alarm and was not under control until 04:33 hours. Whenever I read about fires that start out with the IC thinking single alarm fire for twenty minutes, and it ends a total loss, you have to wonder if a rapid second and third alarms could have made a difference and controlled the fire sooner. It is hard to say for sure, but the delay doesn't appear to have helped much. Failing to call for help early is not unusual, and suggests that some chiefs and departments are waiting too long to call for help.

Don't Dribble In Help

Dribbling in help shows that the IC doesn't really understand some of the basic principles of command.

Dribbling in help is also known as the nickel and dime approach. Instead of getting a lot of help fast, the IC gets a little help now and a little help later, dragging it out over a considerable time. With each call, the IC is hoping that one more company will be all that is needed to control the situation. Firegrounds are unforgiving, and the results of a delay can be very serious. Dribbling in help shows that the IC doesn't really understand some of the basic principles of command.

The nickel and dime approach is a very common fireground problem in fire departments of all sizes. Next time you read a fire article, look closely at the times that help was requested and what was called for, as it can be very revealing. Check out times in the fire log below:

1307 First alarm
1316 Call for one more engine company
1320 Call for one more truck company
1330 Call for one more engine and truck
1335 Second Alarm
1350 Third Alarm
1400 Fourth Alarm

Looking at these times and requests for help, you can bet the farm that fire conditions were marginal until shortly after 1330, after which all hell broke loose. You can also bet that the IC was operating without a net (no staging area) for the first forty minutes of the fire. What effect do you think that had on the safety of the firefighters operating at the scene?

There is usually not a whole lot of building left after a fourth alarm. If the second alarm was called for promptly, could the outcome have been different? It is certainly possible. The IC seemed to think the place was savable early in the incident, because he apparently thought that just one more engine company would control the fire. It seems reasonable to conclude that calling for help quickly and in force may very well have affected the outcome. It certainly would have improved firefighter safety.

I was discussing the problems of the nickel and dime approach with the operations chief of a large western metropolitan department. He said that "we used to have a big problem with it, but not much any more." When I asked him why not, he said "Easy. If you call for a company or two, it had better be the last help you call for." If more help is needed after the initial call for help, the IC is called downtown to explain why he misjudged the fireground situation.

The IC's philosophy should be: "What do I need to handle this incident properly?" not "What can I get by with?" Any dope can call for help when the building is fully involved. A good commander will have adequate help on the way at marginal situations.

WHEN IN DOUBT, CALL FOR THIS

NOT THIS

Calling For Help Because Of The Potential

Often the IC is programmed to call for help based on what is happening, not what could happen.

In the middle of the night, a first alarm response was dispatched for an overturned and leaking gasoline tank truck in an area of garden apartments and single family homes. Initial efforts were made to evacuate the area and call for foam. After the fire department was on the scene for about eight minutes, an occupied house across the street exploded, and the house next door erupted in fire. Second and third alarms were requested immediately.

At this point, calling for help was obvious, but why was help not called earlier? Simple: there was no fire, and we generally have little or no experience with this type of incident. Often, the IC is programmed to call for help based on what is happening, not what could happen. It is important for the IC to anticipate, be proactive and call for help based on the incident's potential, not just what the situation appears to be.

Help with Accidents and Haz Mat Incidents

The principles of calling for help are not limited to fires.

If it's not on fire you probably don't need help. Not true, but too often that's the perception in the fire service. The principles of calling for help are not limited to fires. Whenever any response, such as accidents or hazardous materials incidents, starts to become complicated, call for help quickly and in force. This is much better than calling for help slowly and piecemeal.

The principle of calling for fire companies in blocks, instead of one at a time, also applies to medical units. Many fire departments use a system of medical groups. Each medical group typically consists of four medical units and a medical supervisor.   When fires or accidents result in multiple casualties, the IC can call for medical help in a group. This is similar to the multiple alarm system, and is better then the repeated "send one more ambulance" requests that are so common.

A TYPICAL MEDICAL GROUP

At a mass casualty incidents, the IC can call for a multiple medical group response, requesting, for example, the response of four medical groups. This system makes it easier for the IC, reduces radio transmissions and avoids dumb excited calls for "all available ambulances," which are invitations to chaos.

🔑 When in doubt, call for help quickly and in force.

Calling For Help And Staging Areas

Calling for help without a staging area is inviting the freelance system.

Calls for help should direct the help companies to a staging area, for example: *Command to Dispatch, send a second alarm to stage at Broad and Oak Streets.* Dispatch should notify Command which companies are coming so that the incoming companies can be accounted for and assigned to sectors.

Without staging, the second alarm companies will head right for the incident. At a major rail disaster, a responding second alarm company asked Dispatch: *Where is the staging area?* Dispatch replied: *There is none, we think that they want everybody to come on in.* What effect do you think that this had on the command and control of this incident? Calling for help without a staging area is inviting the freelance system. It says that Command has lost it, so come on in and do whatever you want. Major incidents are tough enough for the IC to handle; the freelance system makes it impossible.

A bonus of staging is that it reduces radio traffic, since all responding companies should stay off the air, leaving the Staging Officer as the only one who talks on the radio. In many departments, the officer of the first company to arrive in staging becomes the Staging Officer. Staging can also reduce the problem of blocked streets and "locked and left" fire department vehicles that often clog major emergency scenes.

The staging area should be located so that responding companies do not have to pass through or around the incident to get to staging. Try to choose a major intersection or parking lot that can handle the companies. During prolonged operations, consider establishing the staging area in a firehouse or community facility. This will give shelter and comfort for the troops and make it easier to pass on important information about the operation.

The staging area should be far enough from the incident to avoid tempting the troops to self commit. I learned this when we called in an engine from staging to discover that they were all dripping wet. When I asked them if it was raining in the staging area, they said nothing but looked sheepish. They were staged too close and didn't wait to be called into action.

During major incidents, consider keeping a minimum number of companies in staging until the incident is controlled. For example, the IC can keep a minimum of two engines and one truck company in the staging area. The IC can give permission for the Staging Officer to work directly with Dispatch to maintain a minimum number of companies in the staging area. This will help the IC, who no longer has to worry about running out of companies. At one recent escalating incident, the Staging Officer asked Command for permission to keep the staging area full. Command, who was up to their ears in problems, said no thanks. Later they called staging for more companies only to find it empty. Guess what! Command then gave permission to the staging officer to keep the staging area full.

**CONSIDER MAINTAINING A MINIMUM COMPLEMENT IN STAGING
DURING DOUBTFUL OPERATIONS**

COMMAND BASICS

Thinking big is hard to do when most of our calls are minor.

A basic in our business is that you set the stage for the big operations on the routine calls. The IC should make routine incidents a command training ground by operating at every incident as if it was a real fire. Upon arrival, the IC should get a good position and set up a fixed command post. Use a command chart to track the sectors and companies. Establish good communications to gather information. Double check when necessary and assume nothing. Think and look at the big picture. The IC should be geared toward major incidents, not small ones, and be able to recognize early on when normal operations won't work or are dangerous. Thinking big is hard to do when most of our calls are minor.

Chief Dinosaur asks, "Why make a mountain out of a molehill and look like a fool?" What Dinosaur is missing is that the troops operating under an alert IC will be ready for the mountains, not just molehills. Dinosaur is right about one thing though; the troops may snicker about your heads-up operation until they realize that it works.

By being prepared, the on-the-ball IC will operate at a working fire almost routinely because everyone is used to the command system. This is because everyone uses the SOPs, the Incident Command System and the Personnel Accountability System regularly. Do not use one operation for "routine" fires and another for "major" fires. By now, you are probably seeing a recurring theme throughout this book: **be consistent and take every call seriously.**

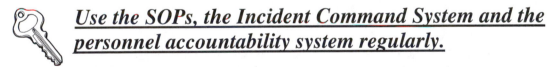

Use the SOPs, the Incident Command System and the personnel accountability system regularly.

Good Chiefs Play The "What If" Game

Mental rehearsal keeps the IC sharp.

When the IC thinks beyond what an incident appears to be into what could happen, the IC is playing the "what if" game. It's easy for the IC at an incident to focus attention on what seems to be happening, rather than what might happen. However, the IC must consider the possibilities. If there is a working basement fire, the IC playing the "what if" game, considers what if the fire extends to the upper floors. This should lead the IC to protect the floors above

194

the fire and to maintain an adequate reserve in staging. The wise commander thinks ahead, covers all the angles, and does not assume that the worst has already happened.

You can also play the "what if" game after the incident to consider what could have happened. For example, following a response to a reported gasoline tank truck accident that turns out to involve a milk truck, the lazy commander thinks no more about it. The alert commander thinks we were lucky, and ponders what should have been done if it was the real thing. The alert commander looks over the milk truck accident and mentally walks through the incident as if it had been a gasoline tanker leaking or on fire. The IC then makes pretend decisions, calls for help, positions companies and sets up sectors as if the situation were real. It may sound goofy, but it works. The alert commander who plays "what if " will do a better job at real incidents because mental rehearsal keeps the IC sharp. One reason that the "what if" game works is that there are often not enough serious incidents to keep an IC sharp. Nobody runs enough gasoline tank truck incidents to maintain the command skills needed. **Fireground commanders have to work at staying sharp.**

The Fate Of The Lazy Incident Commander

Most of command is mental. Most incidents are of little consequence so it's very easy to get mentally lazy. The worst thing that an IC can do is to have two modes of operations, one for ordinary incidents, and the other for working incidents. On ordinary incidents, the lazy IC uses no chart or radio communications because he really doesn't need them. The IC mistakenly believes it's pointless to chart or talk to companies who are verifying a pot of food on the stove.

The IC more or less hangs out waiting for a minor problem to be confirmed. Sometimes, the IC isn't even sure which companies have responded. When this IC tries to command a working incident, it isn't pretty. The IC is not used to working with charts or check-off systems and sectors. Suddenly all the laxness from the routine alarms comes back to haunt the IC, and Command starts to become unraveled.

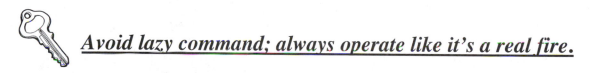

Avoid lazy command; always operate like it's a real fire.

Why Some IC's Scream And Yell

When IC's scream and yell, it is usually because they are so overloaded that they lose their cool. When this happens, important things may be overlooked and safety may be compromised. The better commanders are cool and efficient. The reason they are calm is because they are operating within a system that keeps them from becoming overloaded. The system consists of being prepared all the time, using ICS and SOPs and by practicing management by exception.

For The IC With A Big Ego, Read This

A good engine company can make a mediocre IC look great.

When things go well at a working fire, don't get too smug about it. Consider that the incident may have worked well because a good aggressive engine crew went in and put out the fire, ending all the problems, no thanks to the IC. Take credit where credit is due, but remember that a good engine company can make a mediocre IC look great. God bless good engine crews.

Time Flies When You're Having Fun

What seems like fifteen minutes on a fast moving fireground can be an hour.

It is difficult to keep track of time at an emergency scene. Time can pass very quickly, and what seems like fifteen minutes on a fast moving fireground can be an hour. It is important that the IC keep track of the elapsed time at all working incidents for strategic purposes, SCBA bottle replacement and relief.

Dispatch should announce elapsed time at regular intervals, such as: *"Dispatch to Oak Street Command, you are now ten minutes into the incident."* The time is repeated every ten minutes until the incident is controlled. This delegates the time-keeping task to Dispatch, giving the very busy incident commander one less thing to worry about. A timer can also be used at the command post as a backup.

KEEP TRACK OF TIME

All Companies Are Not Created Equal

The wise IC assigns the right company to the job.

The IC should match the task to the company's expertise. In theory, all fire companies are the same, but, in the real world, some are better than others. In career departments some weaker companies are known by derogatory names, such as "the waterless wonders" or the "the turkey farm." In volunteer departments, the strength of a crew usually depends on who is around. The IC can't do much about what crews are there, but the IC does decide what the crews are assigned to do. Given a choice, the wise IC assigns the right company to the right job. An easy job may be protecting an exposure, and a tough job may be working above the fire in high heat and smoke. If a company is needed to do a particularly dangerous job, a strong company should be assigned to do it, if possible. If a shaky officer or crew is already in a tough position, send experienced and steady people to help them as soon as possible.

Don't Play Musical Chairs With Chiefs

Avoid having the IC always playing startup and trying to figure out what's going on.

The IC at an escalating incident is in the hot seat. It is common for there to be three or four changes of command during the early stages of an incident as senior officers arrive. A big mistake that many departments make is sending the former IC, who had a handle on the situation, off to command a sector someplace. The new IC and each successive IC must start from scratch. The new IC takes time to get up to speed; meanwhile, command suffers. When another higher ranking chief arrives, the same thing happens again. There is no firm, steady hand on the wheel because everybody is always trying to play catch up, and the IC is the lightning rod for everything that is happening. This turnstile command rotation is stressful for the commanders and results in poor fireground management.

This system is simply not a good or efficient way to do business. This happens because many chiefs would rather be leading troops into battle than helping at the command post. Another reason is that some senior chiefs really like to take over and send the former IC away, because it shows everybody who the boss is. Still another reason is tradition, because we have always done it this way.

Changing Command The Right Way

The IC can truly be the commander by using team chiefing.

The first arriving chief officer assumes command. When a higher ranking chief arrives, the senior officer is briefed, and command is exchanged. The former IC now becomes the Operations Officer and continues on with the radio and charts. Team chiefing avoids the chaos that occurs when the former IC who is on top of things leaves to command a sector. The new IC can get some breathing room to think and get a handle on the situation. The incident now has a command team, the former IC and the new IC working together to manage the incident. Team chiefing allows the IC to truly be the commander. "I did this at my first high-rise fire and it worked well. Two heads were better than one. Later arriving chiefs were moved into key sectors." (Battalion Chief David Rohr, (Fairfax County Fire and Rescue, Fairfax, Virginia.)

The IC should be able to sit back and look at the big picture, the overall strategy and tactics. The progressive IC, with a command team can think ahead and be proactive. Compare this to the "Lone Ranger" IC who often simply reacts to what the fire is doing, and is not able to think ahead.

If another senior chief arrives, both chiefs remain at the command post after the new IC assumes command. There are now three chief officers at the command post working together to properly manage the incident. The senior chief is also showing confidence in the junior chiefs, and is developing their command abilities.

Command charts are important in the transfer of command process. One picture is worth a thousand words. The oncoming commander can look at the chart and get a clear picture of the companies on the scene and their deployment. The check-off system shows what has been accomplished, and what remains to be completed. Using a simple written plan as shown earlier in this chapter can also be helpful. This shows the new commander the fireground risks and the goals of the operation.

Use team chiefing to manage escalating incidents.

Assuming Command Of A Messed Up Operation

It is much easier to start at the beginning of an operation and set it up right, than it is to come in the middle of somebody else's out of control mess. Often, the first tip off that it's a lousy operation is when you can't find the IC, or the IC's eyes are spinning.

One fire that stands out in my mind was a working fire in a very large apartment building. Before taking command as the deputy chief, I met with the acting battalion chief (an inexperienced captain) and started asking him questions normally asked at exchange of command:

197

What floor is the fire on? *"Not sure, I think it's on the fifth floor."*

What companies are in fire attack? *"I'll check on that."*

What are the conditions on the upper floors? *"I'll find out."*

What it really meant was the IC was clueless and didn't know what was happening. Later he said, *"I didn't have time to get organized."*

Getting a poor operation back on track is very difficult. You need to have an aide and an Operations Officer to help you. You must organize or reorganize the sectors, find out where all the companies are and see that the proper strategy is being used.

It is also a lousy operation when the IC is organized, but doing the wrong things. An example is an IC who is using a well-organized interior attack for a fire in the attic trusses of a church. In this case, the strategy is wrong.

There is a strong temptation for the incoming IC to stick with the original strategy, even when it is wrong. I'm not sure why this happens; it must have something to do with human nature. One of the most important jobs of the incoming IC is to look objectively at what is happening to determine whether the strategy is correct. Size-up the situation as if you were the first officer on the scene. Determine the objective — what are you trying to accomplish? Review operations taking place and then take appropriate action. The IC must be careful to avoid rubber stamping whatever operation happens to be underway upon his arrival.

One afternoon, I assumed command of a hazardous material incident with fire that involved some serious safety and evacuation problems. Upon assuming command, I told the Operations Officer (who was the former IC and a friend) that safety and control was lacking and needed to be addressed. Nothing seemed to happen, so again I said, "We really need to address safety and control. It isn't working the way it should." The command post was very busy, and he was very involved with other things. It was obvious that I wasn't getting through to him. Frustrated, I finally barked "Listen up, damn it!" There was a stunned silence in the command post. Now that I had everyone's attention, I told them that safety and evacuation were lousy, and I wanted something done about it right now. My actions moved safety and control onto the front burner, and improvements were quickly made. I didn't make any friends that day, but I got the job done. The Ops Officer was cool to me for a while after that; I probably hurt his feelings and embarrassed him. I tried to get it done the easy smooth way, but it hadn't worked, so I had to resort to the iron fist approach. Popularity comes after safety. People get over hurt feelings, but they don't get over being dead. The reason that safety and evacuation were slow was that the CP became so busy that some important things were being overlooked. The job of the IC is to look at the big picture and set the overall strategy.

COMMON SENSE RULES FOR THE INCIDENT COMMANDER

The incident commander has a tough job, and the following common sense suggestions can help to make it a little easier.

The IC Should Not Become The Company Officer

The IC's job is to look at the big picture and make command decisions.

The IC should normally not have to tell an engine company to use a 200-foot attack line or raise a 24-foot ladder. Those decisions belong to the company officer. The department's SOPs should

198

cover most routine situations. The IC's job is to look at the big picture and make command decisions.

Be An IC, Not A Radio Talk Show Host

Getting off the radio allows the IC time to think and plan.

At working incidents, the radio gets busy fast and it's easy for the IC to get married to the radio, and become a radio operator. The more time spent talking, the less time spent thinking, observing and managing. Command must lead, not just talk. The best way for the IC to avoid being buried in radio traffic is to use sectoring, which keeps the radio traffic manageable. By using ICS, the IC has a span of control of about five people.

At major incidents, an aide or an Operations Officer should be assigned to handle the radio for the IC. It's difficult to get off the radio when incidents escalate because it seems natural to continue with what you usually do. Some IC's can't imagine commanding an incident without being on the radio, but getting off the radio allows the IC time to think, plan and confer without constant interruptions. The aide can handle routine messages and refer the important ones to the IC. This is a difficult idea for some IC's to accept, but, at large incidents, the IC can't handle the radio and be the commander. **At working fires you must decide whether to be a commander or radio operator.**

The IC Should Not Become The Water Supply Officer

The IC should designate a Water Supply Officer, not become one.

Sometimes on the fireground radio channels you can hear Incident Commanders talking a lot about water supply. When Command directs companies to a hydrant at 1st and Oak, or to lay a line to Engine 2, the IC is too involved in the details of water supply. The IC's focus should be on the big picture and not one single issue such as water supply. Water supply should normally be handled by company officers and SOPs. If there is a water supply problem, the IC should designate a Water Supply Officer, not become one.

The IC Should Not Become A Reporter

The IC should appoint an Information Officer to handle this task.

An incident doesn't have to be big to be a media event. A school bus accident with a few injuries (while the rest of the kids are looking for a day off) will probably be overwhelmed by TV news cameras. At big incidents, the same thing happens. The media must receive basic information, but the IC doesn't have the time to provide it. The IC should appoint an Information Officer to handle this task. Remember that your incident will make the 6 o'clock news, with or without your help. If the fire department doesn't talk to the media, someone else will. They may interview the old man across the street who will relate that it took the fire department an hour to get there. Be smart and make sure that it is your story that gets reported.

The IC Should Not Become A Rep For Other Agencies

The IC should designate a liaison officer.

At major incidents, the IC is often swamped by representatives from other agencies. The IC should appoint a Liaison Officer to work with and assist these other agencies. It is usually better to conduct a brief meeting with all the involved agencies than to deal with them individually. The IC sets the tone by chairing the meeting, diplomatically making it clear that the other agencies are there to help the fire department. You'd be surprised (or maybe you wouldn't) how many people are IC wannabes. At this meeting, the ground rules are established, and everybody can get a feel for the agencies and resources present. Ask each agency what they can to do to help the operation. Discuss how the different agencies can coordinate with the fire department. Following the meeting, the Liaison Officer handles the details of coordination with outside agencies.

The IC Should Not Become a Traffic Cop

Too often the IC compromises scene safety to accommodate police and transportation personnel.

At incidents where roadways or train traffic are closed, there is a great deal of pressure on the IC to re-open the road or track. It is the responsibility of police and railroads to keep things moving, and the responsibility of the IC to ensure scene safety. While we can't be indifferent to traffic concerns, too often the IC compromises scene safety to accommodate police and transportation personnel. Think twice before making a decision and talk to your Safety Officer and Sector Officers. Err on the side of caution. Consider a partial reopening if safety allows. If allowing traffic will compromise safety, don't do it. **The moral is: Safety first, traffic second.**

Avoid Becoming An Alarm Company Employee

The alarm company, not the fire department is paid to keep up the system.

People pay for alarm companies to service and maintain alarm systems. If we are at the scene of an accidental alarm, the chances are that the alarm company isn't doing the job properly. The job of the fire department is to respond, look for fire and reset the system if it can be done easily, (for example pushing a reset button.) The alarm company, not the fire department is paid to keep up the system. If the system can't be reset easily, notify the alarm company to come and reset it. If possible, notify a responsible party such as a building security officer or building manager of the situation.

There was a chief who kept responding to the same bank after business hours, because of a defective fire alarm. The chief paper-worked the problem in an attempt to get it fixed, but the calls kept coming. The next time it happened he had the troops force entry into the bank to investigate for possible fire. His method was extreme, but they never ran another defective alarm there again. **The moral is: When all else fails, getting mad sometimes helps.**

COMMON COMMAND ERRORS

- Not having a well thought out plan or a backup plan. (I didn't have time to plan.)

- Underestimating the fire potential. (I wonder what happened?)

- Failure to sector, use SOPs and proper ICS. (I usually don't need them.)

- Failure to protect the exposures. (How did the fire get next door?)

- Talking to individual companies instead of sectors. (But it works okay on minor calls.)

- Not having a working fireground radio network. (What radio network?)

- Running around with a portable radio. (I have to see what's going on.)

- Trying to manage a big incident alone. (I'm used to working alone.)

- Failing to call for help quickly and in force. (But I usually get away with it.)

200

- Not recognizing when the troops are in a very dangerous situation. (What situation?)

- Failure to use team chiefing to manage incidents. (But I want to be the big boss.)

- Combining inside and outside attacks. (I didn't know what was happening.)

- Continually rotating offense and defense like a turnstile, where everybody goes inside, and then everybody gets called out. (We've always done it this way.)

- Failure to rehab and relieve the troops. (We've always done it this way.)

Avoid making these mistakes by using a good management system at all incidents.

SUMMARY

Size-up identifies the problems and sets the tone for the incident. The IC must quickly develop a plan based on the size-up.

Plans should be simple.

Never assume that your plan will work; have an alternate plan.

Plans should be flexible and updated by observation and feedback from the sectors.

A stationary command post is important for command and control. The command post should have control and check-off charts to keep track of sectors and companies.

The IC should operate at every incident as if it were a real fire, making routine incidents into a command training ground.

Command should be geared toward major incidents.

Command must recognize dangerous situations, and reduce hazards by avoiding the situation or taking action to make the situation less dangerous.

Recognize that, if you are not operating at your everyday, ordinary residential fire, you are three to five times more likely to die in the building. Fires in commercial buildings have a much higher firefighter death rate than do residences.

A Safety Officer and a rapid intervention team should be used at working incidents.

The philosophy of the IC should be to have enough companies to accomplish the task, and to have a full staging area for unforeseen events.

Using SOPs and management by exception, directions from the IC are needed only when the circumstances are not routine.

The IC must use the ICS to delegate jobs in order to properly command an incident.

At large incidents it is important for the IC to have an aide or aides to help manage the incident and provide information.

Diagrams are needed at the command post during working incidents.

The IC needs to communicate with Sector Officers to stay on top of what is happening.

When major decisions are made, they should be announced to the troops.

Actions reported by the sectors are checked off to show which tasks are completed and what remains to be done. This ensures that all the important points in the operation are consistently and promptly covered.

An easy way to keep track of time is to have Dispatch announce the elapsed time at regular intervals.

14

THE INCIDENT COMMAND SYSTEM

"Control the incident, or the incident will control you." – Battalion Chief Ed Rinker

In the "old days" before ICS, I recall a working fire in an occupied five-story apartment building. I was the IC, with an initial response of four engines, two trucks and a heavy rescue squad, for a total of seven companies. (Remember that the best span of control is about five.) I had to talk to all seven companies. Because of the rapidly deteriorating fire conditions, I quickly called for a second alarm, bringing seven more companies, increasing my span of control to fourteen companies.

The exposure company on the floor above the fire reported that the fire had extended to their floor and they needed another company to control it. At the same time, the second alarm companies were arriving at staging and needed to be assigned. Meanwhile the fireground channel was going absolutely nuts with all the messages. I was simply overwhelmed. In the end, it worked out. The fire was extinguished and the extension to the floor above was halted, but it was due more to the gallant efforts of some experienced companies than a tribute to my commanding.

This wasn't the first big fire where this had happened to me. After every major fire, I would think to myself, there must be a better way. In time, I found a better way—the Incident Command System. Had I used it on this fire, my span of control on the first alarm would have been four instead of seven. After the second alarm, it would have been five instead of fourteen. In short, ICS would have resulted in a safer and more efficient operation.

The ICS can be used on any type of incident by any size fire department. It is not just big-city stuff. There is not just one way to use ICS, it is flexible according to the type and size of the incident. The following information is streetwise ICS, not textbook ICS. The goal is to make your incidents work better.

Use ICS to control the incident, or the incident will control you.

ICS IS NOT STAR WARS

ICS simply provides an organized approach so that everything is covered promptly.

There is very little in ICS that is new to most fire departments. For example, all departments do logistics work occasionally, such as refueling, providing food and drink or replacing air bottles on the fireground. Without an ICS system, we often start thinking about logistics when we start to run out of something. The ICS system simply provides an organized approach so that everything is covered promptly.

ICS Requires Common Sense

While ICS is important, we must put first things first. It doesn't make sense to set up a Logistics Sector when you have people hanging out of windows. If you can quickly extinguish the fire, you may eliminate the need for a big ICS operation. When you have an IC, you have the beginning of ICS, and the rest of the system is set up as time, common sense and resources allow. Given a choice between a chart and a hose line, put out the fire.

Take ICS Seriously

When incidents start to escalate and ICS is not being used, command and control will be terrible.

Some fire service people don't fully believe in ICS. This can include the brass as well as the troops. Oh, they won't admit it, but they consider ICS a little silly, the invention of a goofy commander with crayons and vests. They believe that a real commander needs no props. The problem is, they have not used ICS enough to recognize that the "props" are tools that must be used routinely to organize and to minimize surprises on the fireground.

ICS is a good system, but problems occur when the system is not used correctly. The fire department that doesn't use ICS regularly will have problems, because throwing a little bit of ICS into a messed up operation isn't going to help much. "We never use ICS properly; we piecemeal a little here and there, unless we are taking an assessment center for promotion." (Lt. Jimmy "Hot Shot" Sloan.)

Probably the reasons for this are that that most incidents are simple and really don't need the ICS system. As in, "This is a BS fire. Let's just put it out, mop up and go home." Firefighters are no-nonsense people who often have little tolerance for doing something that they don't view as necessary.

When you don't use ICS, most ordinary incidents will still go all right, but the chances of screw-ups are much greater. When incidents start to escalate and ICS is not being used, command and control will be terrible. When command and control are terrible, it is very easy for people to get hurt. We need to teach ICS, preach ICS and use ICS regularly because the Incident Command System is a vital management tool.

Use It Early And Use It Often

Most company officers are not as familiar with ICS as they should be. They often say that they have a handful of fires yearly and most are quickly knocked down, so that they never really get a chance to use ICS. This can occur because some IC's use two modes of operation. On minor incidents, ICS is not used because as one chief said, "It's silly to use all those ICS terms and stuff on minor incidents. When I have a working incident, then I start ICS." This approach, doesn't work very well. The company officers won't really know ICS because it isn't used often enough to become part of the regular fireground routine. To be effective, ICS must be used routinely.

SECTORING AND HOW IT WORKS

All companies should be operating in sectors at all multi-company incidents.

Sectoring is **very important** because the IC can be easily overwhelmed as an incident escalates, causing a communications overload, and a loss of control. Sectoring divides the incident into manageable pieces so that the IC can maintain an overview of operations without having to concentrate on any one segment. The maximum number of people reporting to any one person should be about five. Delegating authority through sectoring also shows the Sector Officers that the IC has confidence in them.

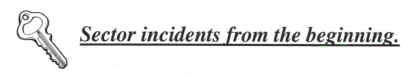

Sector incidents from the beginning.

ICS must be used at every incident requiring more than one company. This means that it will be used a lot and everyone will become familiar with it. When the big one hits, companies will be prepared because ICS will be part of their everyday routine. It is easiest to assign companies to sectors when they are put to work. For example: *Engine 9 cover exposure B, you are working for the Exposures Sector.* Communications are then between the IC and the Exposure Sector Officer, not between Engine 9 and the IC.

Use ICS at all multi-company responses.

Sector Leaders

Choose experienced and reliable Sector Officers.

Each sector is assigned an officer who directs, coordinates and ensures the safety of personnel in the sector. As an incident escalates, chief officers should replace company officers as Sector Officers to allow the company officers to command their companies. For example, a battalion chief will do a better job as the Vent Sector officer than a company officer wearing two hats, commanding Engine 2 and in charge of Vent Sector. Chief officers should be assigned quickly to the most critical sectors. Usually the most important sectors are: suppression, exposures, medical, water supply, safety, search and rescue and haz mat. Normally, only a few of these are really critical at any one incident.

The IC must delegate to sector leaders.

A SECTOR LEADER

Sector Officers must be aware of the companies assigned to them as well as the strategic plan. For example, when Chief 5 is assigned to the Interior Sector, he is told whom to relieve and given a list of the companies in the sector, their functions and the sector radio channel. Sector Officers keep the IC informed about conditions in their sectors. Sector Officers should use small command boards to keep track of their sectors. In complex situations, Sector Officers can benefit by having an aide to assist them in managing the sector. A sector chart sample follows:

EXPOSURES SECTOR CHART

E 16 THIRD FLOOR

E 2 EXPOSURE 2

T3 EXPOSURE 2

T1 THIRD FLOOR

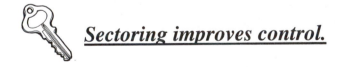 *Sectoring improves control.*

Sector Officers also maintain unity of command, so that everyone reports to just one person. This avoids firefighters receiving conflicting orders from different officers, a problem that almost never happens. Just kidding. I wanted to see if you were paying attention.

SECTOR LEADER

SECTOR LEADERS MANAGE THEIR SECTORS

Why Titles Are Important

If Engine 1's officer is assigned to lead the Exposures Sector, the proper radio designation should be "Exposures," not "Engine 1," or "Billy." This allows everybody on the fireground to be able to relate to the messages to and from the Exposures Sector. Messages from Engine 1 mean little unless you know their assignment. If you are searching the 2nd floor and you hear Engine 1 screaming that the fire is getting away from them, it's really important for you to know where they are. Are they are on the floor under you, or in an exposure someplace? That is "must know" information. Another reason to use proper titles is if Engine 1's officer is relieved by another officer, the radio designation doesn't change. It remains "Exposures."

House Fire ICS

The chart below shows ICS at a typical working house fire. When additional companies are assigned to a sector, the IC's span of control usually does not increase.

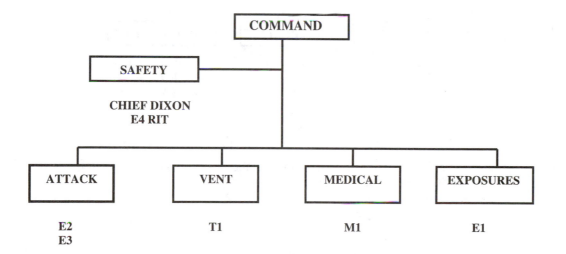

The Same House Fire Becomes A Third Alarm

Assume that this fire becomes a third alarm, because the fire is in a very large house with exposures being threatened. The third alarm almost triples the number of companies, but Command's span of control changes very little. Implementing an Operations Officer keeps Commands' span of control manageable.

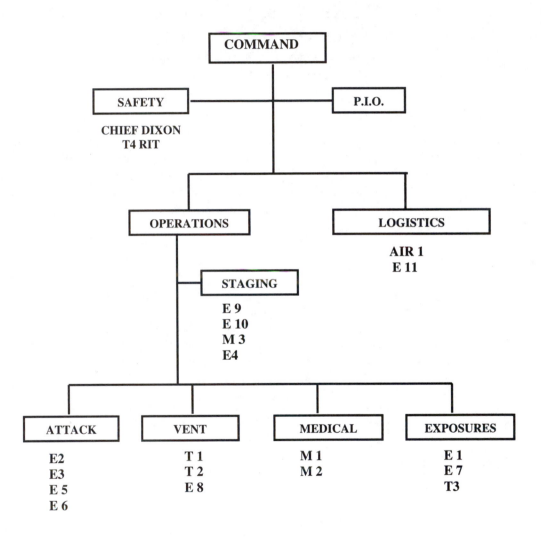

This chart demonstrates the beauty of the ICS system in a nutshell. As the incident expands, the span of control remains about the same.

Really Major Big Time ICS

The following chart shows using an Operations Officer and Branch Leaders at a really big operation. When the number of Sector Officers exceeded five, the Operations Officer assigned Branch Leaders to take over supervision of some of the sectors. Assuming that each sector contains four companies, there are a total of 64 fire companies at this scene. Notice that nobody's span of control exceeds five. What a great system!

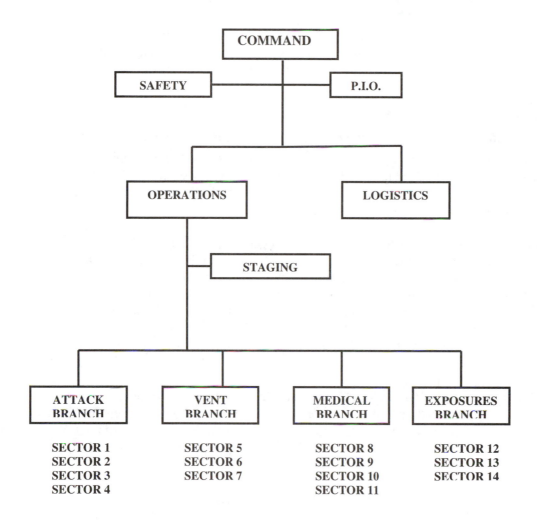

The Planning Officer

At complex or escalating incidents, a Planning Officer should be assigned. This is probably the most neglected position in the ICS system. The Planning Officer is the "what if" person who has the job of looking ahead and forecasting what may happen. It is easy for the IC to get totally absorbed in what is happening, and not think about what could or might happen. For example, the Planning Officer may look at a fire and forecast "what if" the fire extends to the exposures, or "what if" the haz mat leak ignites. The Planning Officer then advises the IC of the possibilities and recommends preventive measures to be considered and additional resources needed if the possibilities do occur. This will help avoid playing catch up later.

HOW TO SCREW UP ICS

Several areas of ICS can become problems if personnel do not train and manage smaller incidents using their department's incident command system. These areas include span of control, assigning sectors, delayed implementation and taking and passing command properly.

Exceeding Span Of Control

On paper, many departments are committed to ICS, but the reality is something else.

I just finished listening to the fireground channel at a two-alarm house fire involving about a dozen companies. There was absolutely no ICS used. The IC talked to each company individually. No staging area was used. All the second alarm companies were told to report to the command post.

Another example of a poor ICS operation was a fire that occurred not long ago in a townhouse. The initial response was three engines, two trucks, an ambulance and a battalion chief. The first alarm companies were not sectored, but one of the engines was called "Interior." The IC soon called for help: two more engines, a truck company, an air truck and another battalion chief. When they arrived, the IC directed the first engine to the adjoining townhouse on the left side (Exposure B.) The second engine was directed to the attached exposure on the right (Exposure D.) The IC did not designate an Exposure Sector Officer, so both engines reported to the IC.

The additional truck company was ordered to the rear. The rear was not a sector, so I'm not sure what they did, and I doubt that the IC knew either. The other chief was also sent to the rear, but not assigned a sector or companies. The result was that the IC had a span of control of ten, twice what it should be, and the other chief had none. On paper, this department is committed to the ICS, but the reality is something else.

Unfortunately, this fire is probably typical of ICS in some fire departments. They use a few ICS terms, but really don't use ICS. Too many departments don't know any better; to them poor ICS is normal. This fire was eventually extinguished, but if the fire had decided to head down the block, it would have been an even bigger command disaster than it was.

Many fire departments fail to use an Operations Officer and Branch Leaders as incidents expand. This occurs because very few incidents involve more than a handful of sectors. The tendency is for the IC to just keep adding sectors, because that is what he normally does. The result is that that the IC can wind up trying to juggle fifteen sectors at major incidents. This can be avoided by adding another level whenever anyone's span of control exceeds five.

If the IC's span of control greatly exceeds five, things turn ugly real fast.

"Interior Sector" Problems

The IC failed to sector early, so there really was no "interior sector."

During a commercial fire, the IC started out assigning individual companies without assigning sectors. The fire escalated, making it very difficult to reorganize all the companies into sectors. When another chief officer arrived, the IC assigned him to the "Interior Sector." The problem was that there really was no sector, just a bunch of companies that should have been sectored from the beginning. The companies were not aware of being within the "interior sector" and individual companies continued to talk directly to Command. The chief assigned to this sector didn't know what companies were assigned or what they were assigned to do. Because the IC failed to sector early, what really happened was that the IC assigned the chief to report on what was happening inside. A true interior sector should have been set up from the beginning of the operation with a Sector Officer, a reasonable span of control and good communications.

What Rear Sector?

It is a common error for an IC to make assignments to a rear sector during interior fires. The problem is that there is no rear sector on the typical interior operation. Some companies may go into the fire building through the rear, but they are operating in interior sectors. A company that is assigned to the non-existent rear sector is not going to stand around in the back yard at a working fire, so they will probably freelance. If you want to know what is going on in the rear of the structure, assign an aide or an individual firefighter to this specific task.

During defensive outside operations, there is usually a sector operating in the rear. Assigning Sector Officers and companies to the rear sector works well at these defensive fires.

Don't Waste Chief Officers

Chiefs are often wasted as Staging Officers, but it may be a good place for a chief that you want to get rid of. A Staging Officer is really a traffic cop with a clipboard. It is an important job but hardly a command position. During a major interior operation in an office building, a chief was heading the staging area. Over time, firefighters became worn out and required medical treatment. Soon a trickle of exhausted firefighters became a flood. It quickly started to develop into a mini-mass casualty. The Medical Sector was not under the command of a chief officer and was not properly handled. Meanwhile a senior, experienced chief officer was wasted in the staging area. An exception is disaster type operations involving very large numbers of companies.

A working fire in an apartment building involved rescues being made by ladders and fire escapes. There were two sector chiefs, one in charge of the outside Rescue Sector, and the other in charge of water supply. However, there wasn't a serious water supply problem. Instead, a chief should have been in charge of the Fire Attack Sector, because if they had put out the fire, most of the problems would have disappeared.

211

Still another example of wasting a chief is assigning a chief to the non-existent rear sector during interior operations, typically to find out what's happening in the rear. The chief in the rear sector is really being used as an outside observer. It doesn't take a chief to tell you that the fire is coming out of a window. Most firefighters could do that, and some might even do it better. The IC's concern in getting a view from the rear is legitimate, but a chief is the wrong person to do it. SOPs should cover this situation by using an aide or company officer as a rear observer. Chief officers should be assigned to the most important sectors.

The Folly Of Delayed ICS

You sometimes read excerpts from fire articles in the fire magazines such as the following: "The chief of department arrived twenty minutes into the incident, saw the magnitude of the fire and implemented ICS." That is where I stop reading the article. The fireground simply doesn't operate that way. If you have an escalating incident with no ICS being used and believe that you can blow a whistle and change over to an orderly ICS fireground operation, you are into serious unreality.

ICS Principles? What Principles?

For ICS to work properly, it is essential that everyone understands the system and it's basic principles. While listening to a fireground operation, I heard the IC organize three companies into a Ventilation Sector, assigning one company officer as the Sector Officer. The IC then communicated individually with each company in the Vent Sector and never allowed the Sector Officer to run the sector. It seems that this IC was not aware of the ICS principle that sectoring should be used to maintain a proper span of control.

To use the ICS system properly, everyone should be aware of how it works and why it works. A good way to understand ICS better is through table top drills, and to talk it up following your incidents.

ICS Command Problems

Using the ICS, command is usually established upon the arrival of the first officer. Many departments allow the first officer to pass command to the next arriving officer if necessary. Sometimes departments have problems with passing command, as in the following examples:

- The officer who always wants to pass command to be in on the action.

- The officer who always wants to retain command to not be in on the action. (Every department has a few of these.)

- The officer who takes command but does not command. (Every department also has a few of these.)

The following guideline may help your command passing process, while allowing flexibility.

> *If the first arriving company has a working fire and is going to make an interior attack, the officer should pass command and lead the attack. If there is not going to be an interior attack, the officer should generally assume command. When this happens, an acting officer is moved up to be in charge of the company.*

When In Command, Command

The company officer taking command is the IC and should take charge of the operation. Avoid situations where the company officer (who is the IC), reports the situation to the responding chief officer who has not yet arrived. For example:

Engine 3 to the chief, we have food on the stove and can handle it by ourselves.

The chief then makes the final decision and returns the unneeded companies. **Time out!** The company officer is in command. So, Command (Engine 3) should make the decision and report it to Dispatch. If we expect the company officer to function as a command officer, we must let him command. A basic principle is: when in command, command.

THE IMPORTANT THINGS THAT MAKE ICS WORK

ICS has many parts, but the trick is to focus on those that are really important. To make ICS work better in your fire department, consider the following.

Make ICS User Friendly

It is important that your ICS works well in your fire department. Avoid a generic ICS package that was never modified to suit your needs. Throw away the complex charts and thick texts that nobody understands or follows. Make ICS simple and user friendly and then use it. Evaluate and revise your ICS annually to make sure that it is working correctly. The post incident analysis is an excellent way to ensure that the ICS system is operating the way it was designed.

Use ICS All The Time

One of the keys to ICS is to use it all the time, every day, on all multi-company incidents. Although this is emphasized in all the ICS courses and articles, it can be hard to do.

Maintain Control Through Staging

Companies or individuals must report to a staging area or control is lost. It's as simple as that. No staging, no control. Using a staging area improves personnel accountability and reduces freelancing (kamikaze firefighters hate staging areas.)

Safety First

Using a qualified Safety Officer on working incidents can do wonders for the safety of your operation.

Establish Logistics

It is important that logistics be addressed early. If the IC is too busy, logistics should be delegated to a Logistics Officer. Failure to set up logistics is a recurring problem in many departments.

Advance logistics planning pays off. Establish contracts with restaurants, portable toilet companies, medical supply concerns and other businesses for the logistical support needed at major incidents. The Los Angeles City Fire Department has developed a resource book that shows pictures of equipment, what the equipment does, equipment location, estimated response time and costs. This resource book has greatly improved their logistical support system. A similar book showing your local resources can improve your operations.

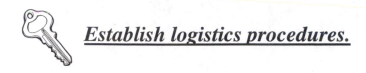

Establish logistics procedures.

Use Vests

Vests should be worn by command personnel to identify their job functions. "We don't need no vests, I already know everybody," said the lieutenant. "That's true," replied the chief, "but the vests identify the **job** to which the person is assigned, not the person." It is very important to be able to quickly identify the key players at an incident, and vests are the best way to do it. Vests must be carried on all front line vehicles, so they will be used promptly and often. Don't keep the vests in storage waiting for a plane to crash, make them part of your everyday operation. Color coding the vests by function, such as red for fire personnel, and yellow for medical, makes identification easier. Vest colors and titles should also be standardized with your neighboring departments.

USE VESTS TO ID THE PLAYERS

Use ICS Cheat Sheets

No one can be thoroughly familiar with job assignments that are infrequently used.

The ICS cheat sheet shows the basics of each ICS function. Each person assigned an ICS position can look over the cheat sheet to review their job duties. Some departments laminate these sheets and keep them in the pocket of the ICS vests. This acknowledges that in the real fire service world no one is thoroughly familiar with job assignments that are infrequently used. The handout eliminates the "seat of your pants" operation where Sector Officers are expected to lead in situations that they have never faced before. A sample cheat sheet for the Logistics Officer follows:

214

```
                    LOGISTICS OFFICER


        OBTAIN SITUATION BRIEFING FROM COMMAND

        DON LOGISTICS VEST & ASSESS SITUATION

        RADIO CHANNEL _____

        RADIO DESIGNATION IS— LOGISTICS

        NORMAL LOGISTIC FUNCTIONS ARE:

        *BREATHING AIR SUPPLY

        *WATER SUPPLY    *FUEL SUPPLY

        *FOOD            *SET UP REHAB

        *OBTAIN RESOURCES NEEDED
        (RESOURCE AVAILABILITY LIST ATTACHED)

        COMMUNICATE WITH COMMAND:

        *YOUR ACTIONS   * RECOMMENDATIONS

        *STATUS OF RESOURCES
```

Enforce ICS

A fire department must use ICS for a long time before it is a regular part of the emergency scene. ICS must be used regularly, because many chiefs and company officers tend to slide back into their old ways. A good way to enforce proper ICS is to refuse to accept incorrect ICS. For example, if Engine 1's officer is assigned to lead the Exposures Sector, but incorrectly calls Command as Engine 1, Command should correct them on the spot, by replying: *Go ahead Exposures, this is Command.* Using this method, it won't take Engine 1 very long to get back into ICS. But if they are uncorrected the problem will get worse, and if you condone it, you own it. **Never reward unacceptable performances or behavior in your ICS, (or anywhere else.)**

Maintain Proper Span Of Control

Always watch your span of control. The ideal span is about five. Whenever the span of control exceeds five, add another level as shown below:

- If a sector starts getting too large, add another sector

- If the IC's span grows to greater than five, add an Operations Officer

- If the number of sectors exceeds the span, add a Branch Leader

The ideal span of control is about five.

Built-In ICS

You can design your own ICS system (remember that ICS is flexible) to make it easier to operate. One way to do this is by pre-arranging your sectors to suit the needs of your department. An example of built-in ICS follows:

- The first engine will assume the Attack Sector at structure fires

- The first truck company assumes the lead of the Vent Sector

- The first medical unit leads the Medical Sector

- On calls for help, the first officer to arrive at the staging area becomes the Staging Officer

- When a senior officer assumes command, the former IC remains at the command post and becomes the Operations Officer

- The second chief officer to arrive becomes the Safety Officer

- The third arriving chief officer handles logistics

Some departments have ICS built in to the point that when an incident reaches third alarm proportions, the entire ICS system is in place without any orders being given. When this system is used, it is important that command officers be able to recognize situations where pre-arranged ICS does not apply. When unusual conditions exist, commanders must adjust to deal with the situation at hand.

SUMMARY

Design your ICS system to meet your needs.

ICS can be used on any type of incident by any size fire department.

There is not just one way to use ICS. It is flexible according to the type and size of the incident.

Use ICS regularly. What we do on the everyday incident becomes a habit.

All companies must be operating in sectors at all multi-company incidents.

Command maintains the proper span of control through sectoring, which divides the incident into manageable pieces. The maximum number of people reporting to any one person is about five.

Each sector is assigned a Sector Officer who directs, coordinates and ensures the safety of personnel in the sector.

Sector Officers must be aware of the companies assigned to them as well as the strategic plan.

Chief officers should be assigned to the most critical sectors.

Some important things that make ICS work are: using it all the time; using staging areas, vests and cheat sheets; staying on top of logistics; and enforcing ICS on a day to day basis.

The post incident analysis is an excellent way to ensure that the ICS system is operating properly.

15

EXPOSURE PROTECTION

"Don't concentrate on where the fire is; look to where it is going." - Chief Edward F. Croker, Fire Department, City of New York

Karl von Clausewitz, the famous student of warfare and philosopher, once remarked that everything in war is simple, but that the simplest thing is very difficult. This statement is equally true in the fire business. Exposure protection is simple and important, but it can be very difficult to do. We have all seen the fire thought to be controlled that breaks out in the attic or an upper floor. One of the reasons that we fail to protect exposures properly is complacency. Since most fires don't require much exposure protection, we can get out of the habit of protecting exposures.

One bright and sunny day, there was a working fire in the cockloft of a townhouse. The firefighters thought that it was knocked down when heavy smoke started to roll out of the attic of the house next door. The troops went to work on the neighboring cockloft, and the same thing happened; the fire moved to the third house. Fortunately there were only three houses in the row. At the rate they were going, if the row of houses had been a mile long, they would probably still be there going from house to house.

Chief Dinosaur says, "Hell, we don't have enough people to put the fire out; we can't protect exposures." Everybody has been in that situation at one time or another, but you do your best to protect exposures. Command should call for help to get the personnel needed for exposure protection. Both internal and external exposures should be protected. If the fire is knocked down before the exposures are covered, don't let the troops stand around patting each other on the back. Quickly check the exposures.

A Frightening Exposure Problem

I was the IC at a serious basement fire in a two-story townhouse that occurred in the middle of the night. The fire was knocked down fairly quickly, and the exposure company reported very heavy smoke in the attached building on the left (Exposure B.) Next they checked the attached building on the right (Exposure D), and they found light smoke. The zinger was that they discovered two elderly bedridden people alone in the exposure on the right side. If we had not checked the exposures quickly, and these people been in the exposure on the heavy smoke side, they probably would have died in the exposure.

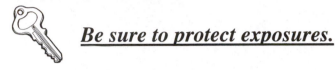 ___**Be sure to protect exposures.**___

They Forgot about Interior Exposures

This department went from a working basement fire to a working high-rise fire in two minutes.

A few years ago, there was a fire in the basement of a large old commercial building. The fire was major, and almost all efforts were directed at finding and extinguishing the basement fire. There were no external exposures, but there were about twenty stories of internal exposures above the fire. Because of the intensity of the basement fire, and the large number of companies needed to combat it, the IC had tunnel vision and did not protect the floors above the fire promptly or thoroughly. When the basement fire was finally controlled, companies were sent to check the upper floors. They found several working fires on upper floors as a result of the extension from the basement fire. This department went from a working basement fire to a working high- rise fire in two minutes. The same thing happens in house fires when a basement fire extends to other floors or the attic, often much to the surprise of the commanders. **The moral is that internal exposures must be protected at all fires.**

FAILING TO PROTECT INTERNAL EXPOSURES

Exposure Priorities

Priority should be given to life hazards. If one of the exposures is an occupied nursing home and the other is an empty store, the decision is easy. Another consideration is property value. If one exposure is occupied and the other is an abandoned house, obviously the occupied house is the priority. Distance to the exposures is another consideration. Protect the closest exposures first. Don't forget that exposures that are higher than the fire can be in more danger than exposures on the same level.

HIGHER EXPOSURES CAN BE IN MORE DANGER

Checking Exposures

Command should designate an Exposure Sector and assign sufficient personnel to accomplish the task. Avoid having somebody "keep an eye on the exposures." Exposure companies are normally assigned to protect the floors above the fire, exposures on the fire floor and exterior exposures. Companies assigned to protect interior exposures in the fire building should be told the room or apartment number and the quadrant where the fire is located. This will allow the exposures to be checked more quickly and accurately.

PROTECT EXTERIOR EXPOSURES

PROTECT EXPOSURES ABOVE THE FIRE

PROTECT EXPOSURES ON THE FIRE FLOOR

EXPOSURE PROTECTION

When protecting exterior exposures, firewalls are a good sign, but don't count on them because they may have openings caused by poor workmanship. In addition, any sprinkler systems in the exposures should be supplied. Sometimes, it is necessary to put water on the face of the exposed buildings, wetting them down to prevent the fire from igniting them. This can be done with hand lines or master streams, depending on the circumstances.

221

When a company is assigned to check exposures, the company should:

- Work within the Exposure Sector

- Check the most important exposure problem first

- Bring a hose line, hook, halligan bar, stepladder and a thermal imaging camera

- Open up walls and pull ceilings, if necessary, to see if the fire has spread

- Give periodic progress reports to Exposures on what is seen and found

- Stay put until relieved

Exposure companies should remain protecting the exposures while keeping a constant watch. It is a good idea for the Sector Officer to closely supervise the exposure companies. Fire companies want to be in on the action and are usually not thrilled to be sent into an exposure. They tend to slide off into the main fire fighting operation (which is what they really wanted to do in the first place) if they find no immediate threat to the exposure. At a store fire, I assigned an engine to the attached exposure building. I never heard from them, so I called them on the radio but got no answer. After repeated calls, they finally answered. I asked them what was going on in the exposure. They answered: *"The exposure was all right a little while ago and we are now on a hose line in the fire building."* I told them: *"Go back where you are supposed to be and to report to the command post before you leave the fireground."* This message also served notice to the other companies that if you don't do what you are supposed to do, you'll get nailed. It could have been worse. They could have snuck out of the fire building to sit in the exposure.

EXPOSURE B   EXPOSURE D

PROTECT THE EXPOSURES
Courtesy of Dennis Wetherhold Jr.

Strip Shopping Centers

If a store fire cannot be quickly knocked down, write it off and quickly protect adjacent stores.

I once observed a fire in a strip mall that started in the grease duct of a Chinese restaurant. There were three or four companies attacking the store fire, but nobody in the stores on either side. At the opposite end of the strip mall was a furniture store that was open for business and was probably unaware of the fire. Before it was over, the clerks and customers were running out of the furniture store in a cloud of smoke. The furniture store was lost, along with an entire block of stores. Lack of exposure protection doomed that shopping center.

At a well-involved store fire with no life hazard, consider using master streams from the beginning, because hand lines are usually inadequate for this type of fire. If you screw around with hand lines and concentrate on the fire (which is very easy to do), you may lose the whole shopping center. If a store fire cannot be quickly knocked down, write it off and cover the adjacent stores.

You must get ahead of the fire. Skip a store or two on each side if necessary and trench cut the roof or pull the ceilings, creating an interior trench cut three feet wide, running the length of the store from the front wall to the back wall. This is where you make your stand with hand lines to cut off the fires spread. Opening the roof will help prevent the fire from going sideways, but avoid working over the fire and beware of truss roof construction and collapse potential.

Anti-conflagration Strategies

It takes a lot of fire companies to extinguish a major fire, but it usually doesn't take many companies to keep a fire from getting into the exposures.

When conflagrations occur, the original strategy was often to extinguish the fire, neglecting the exposures. Avoid getting drawn to the flame, where everybody focuses on the heavy fire. Consider cutting the fire off or writing off the fire building before it's too late to save the whole area. Resist the temptation to put streams wherever you see the fire. **Concentrate on where the fire is going, not just where it is.** If upon arrival, a building is so heavily involved in fire that you cannot reasonably expect to knockdown the fire with your resources, write off the building and concentrate on the exposures. I know, Snyders hardware store is a landmark that has been there for a hundred years, but it may be a choice between losing Snyders, or losing Snyders along with the rest of the block. In extreme cases when the fire is spreading rapidly and your resources are thin, ignore the fire building **and** some of the exposures. The idea is to set up a strong defense in advance of the fire. Avoid playing catch up and chasing the fire down the block. Every effort must be made to get ahead of the fire. Writing off the fire building and some exposures to set up an anti-conflagration position will probably work, but it's very difficult to do at the once in a generation fire.

It takes a lot of fire companies to extinguish a major fire, but it usually doesn't take many companies to keep a fire from getting into the exposures. Usually all it takes is commitment and vigilance. If this was done early in many of the fires that did become conflagrations, the chances are good that the fires would have been controlled at the strong advance defensive points. **The moral is get ahead of the fire; don't chase it.**

SUMMARY

Protecting exposures is very important. Both internal and external exposures should be protected.

Exposures companies should be clued in to the fire location by using apartment or room numbers and the quadrant system.

Companies protecting exposures should use thermal imaging cameras and if necessary, open up walls and pull ceilings to check for fire spread.

If a building is so heavily involved in fire that you cannot reasonably expect to knockdown the fire with your resources, write off the building and concentrate on the exposures.

16

DEFENSIVE OPERATIONS

Hope for the best, but prepare for the worst. - English proverb

A sky full of smoke and fire should tell us to start thinking about a defensive operation. Arriving at the scene of a well-involved warehouse fire, the first engine did what they always did at fires, they stopped in the front and came off with an 1¾-inch attack line. The problem was that the warehouse was beyond saving, and there were severely exposed buildings on both sides. I was with the third arriving engine company and wound up protecting one of the exposures, which was only a dozen feet from the warehouse. It was a tough fight, but, with help from other companies, we were barely able to keep from losing the exposure building. The first engine may as well have stayed back in the firehouse. In fact it would have been better if they had because they were wasting water that was needed for exposure protection.

Why did the first engine screw-up? Because they had the mind-set that what is normally done at most fires is what should be done at all fires. We must remain flexible in our business and not get locked into a "one size fits all" mentality. We use the 1¾ lines so often and so successfully that we tend to overuse them.

A fire in a building under construction involved most of the fifth floor, and the strategy was to use hand lines. The hand line attack took a long time to position, and was unsafe and ineffective. After it failed, the strategy was finally changed to a defensive operation. An aerial tower was used to knockdown the fire in short order. It was obviously a defensive operation from the beginning, but the first engines went up with hand lines, and the IC continued this ineffective strategy until it predictably failed. To avoid these problems, companies and commanders should think master streams at any doubtful operation. A lot of water quickly may work.

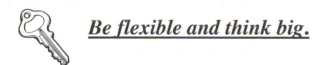 ***Be flexible and think big.***

225

Changing from Offense to Defense

Err on the side of caution. When in doubt, pull them out.

If inside crews are not advancing and making progress, Command should start preparing for master streams. This usually means calling for help, positioning towers and ladder pipes, readying quints, and deck pipes, and considering the water needed to supply them. Too often, we wait until the interior operation fails before we prepare for a master stream operation, creating a big time gap when nothing is being done.

When should the IC change from an offensive to a defensive operation? There is no set formula, but some good rules of thumb follow:

- The structure cannot be saved

- There is doubt about the stability of the building

- The fire is on more than one floor

- There is little to gain, such as in abandoned building fires

- The fire is in the trusses, and cannot be knocked down quickly

- Time is passing, and little progress is being made

If you wait too long, conditions may deteriorate to the point that all the firefighters may not be able to find their way out. Don't wait until the last second to pull out the firefighters. An orderly withdrawal is safer and makes it easier to do an accurate accountability check. When the troops are forced to retreat in haste, it is more difficult to reorganize the forces into effective action. Err on the side of caution; when in doubt pull them out.

Bailing Out

All fire department's need procedures for abandoning fire buildings that are understood by everyone. Some departments' entire evacuation procedure is to blow all the air horns. This may get everybody out, but can cause problems. The air horn doesn't tell you why the building is being abandoned. Does it matter? You bet it does. If there was a partial collapse or other danger near an entrance, blowing the horns may bring the troops out right under the potential collapse. Also, air horns don't tell you the reason for the evacuation. If the building shows signs of collapse, you obviously want everybody really hustling out. If the reason is just to shift into a defensive mode, speed may not be that important.

A good basic evacuation procedure is for Command to notify all sectors and Dispatch, indicating the reason for the bail out. For example:

Command to all sectors—Abandon the building, we are going into a defensive operation.

Command to all sectors—Emergency evacuation, abandon the building quickly, the building is unsafe.

Command to all sectors—Emergency evacuation, abandon the building, do not use the rear exit because of an unstable wall.

226

Sector Officers then notify all companies assigned to their sectors, using face-to-face communications if possible. After the radio announcements have been made and acknowledged, some Dispatch Centers have evacuation tones that can be sounded. Command can also order the sounding of apparatus air horns for ten to fifteen seconds to reinforce the message.

When an emergency evacuation is ordered, all firefighters should exit the structure. Evacuation procedures should consider that an engine company protecting other firefighters, such as a search team, can't just leave. It is important to make sure that the search team gets out before they leave their hose line. Crews must also understand that they should not waste time trying to recover fire department property before retreating from the building.

It is a mistake to assume that Command must order all retreats. Sometimes conditions deteriorate quickly, before Command is fully aware of what is happening. When this happens, it may be necessary for Sector Officers to make the decision to retreat, and then notify Command.

Bailing Out Problems

There is always somebody who thinks they can knockdown the fire if you give them a few more minutes.

Many fire department's have problems with certain firefighters who resist leaving the building when ordered out. You may have a Lieutenant Kamikaze who insists that he can knockdown the fire if you allow a few more minutes.

"All of us who were fighting the attic fire were ordered out of the building. Some did, some just backed down to the second floor, and some had be kicked out of the attic." (Captain Thomas Monroe, Engine 6.) The firefighters who are foot dragging mean well, but are operating with tunnel vision. They are a danger to themselves and others. Probably the best way to handle this problem is to raise hell if it happens and stress the problem in training. **Make it clear that "Get out" means get out.**

 **_Companies must leave when ordered out._**

Accountability

Companies ordered out must promptly leave the building together as a company. This is where sticking together at routine incidents pays off. When companies have exited the building, the company officers immediately take a head count of their crews, reporting the results to their Sector Officers. Do not take the whereabouts of any member for granted. Verify everyone's location. The Sector Officers check to see that all companies in their sector are outside. No defensive operation begins until it is verified that everyone is out of the building.

How to Reorganize into Exterior Sectors

The location of a company's apparatus usually determines their sector assignment.

After the troops are outside, it is hell getting them reorganized into defensive sectors. During inside operations, all the companies are in sectors such as Interior, Search, Vent, and so on.

When changing to master streams, all sectors must be realigned into outside sectors to cover all sides of the building.

Command begins by assigning a Sector Officer to each exterior sector. These Sector Officers then assemble their own sector personnel from the companies with apparatus already within the sector. They report to Command what companies are in their sectors, as well as any overages or shortages of companies needed within their sectors. Command then shifts companies among sectors as needed, and brings in fresh companies from staging. This method is fast and dirty, but it works.

Master Stream Operations

Cut the fire off if possible.

There is an old saying, "When the pipe goes up, the building goes down." This means that when we have to resort to elevated heavy streams, the building is lost. While this is sometimes true, good stops can be made with master streams. A rapid knockdown may keep the structure from being a total loss, and master streams can often save exposures.

GET AHEAD OF THE FIRE
Courtesy of Fairfax County Fire and Rescue, Fairfax, Virginia

Call for towers and truck companies early, and then have an aide find and hold good truck positions. When positioning apparatus for master streams, notice the direction that the fire is moving and position companies to cut off the fire. Consider the time needed to set up master streams and estimate likely fire conditions when they are placed in operation. Bring trucks coming from staging first, followed by the engines used for water supply. For example: *Have Truck 1 position on Oak street, followed by an engine company for their water supply.*

Position companies to cut off the fire.

Use straight streams and move them around as needed. Don't keep streams in one place until the Mississippi river starts to run out the front door. Don't operate streams in areas that are lost if there is anything, anywhere to be saved. When you are going into defensive operations, consider establishing roving brand patrols for firebrands and embers.

Working with the Big Boys — Towers

A well-positioned tower can be your alternate plan.

Aerial towers are much more effective than ladder pipes, deck pipes or monitor nozzles. Towers provide more water, more mobility and more safety. One tower in the right place can equal a third alarm in effectiveness. On any doubtful operation, the presence of a well-positioned tower can be your alternate plan. If you have towers in your department or through mutual aid, make sure that they get to your working fires. Towers are also valuable to protect exposures as shown below.

A TOWER PROTECTS THE EXPOSURE

Don't Use a Firefighter to Direct a Ladder Pipe

A firefighter directing a master stream from the top of an aerial ladder closely resembles a hot dog at the end of a stick.

The only advantage to using a manned ladder pipe is that you may gain a better view, and, maybe, to direct the stream more accurately. The disadvantage is that the firefighter directing a master stream from the top of an aerial ladder closely resembles a hot dog at the end of a stick. The firefighter may fall or be caught in a flashover or explosion. There is also the risk of mechanical failure of the ladder pipe assembly or the ladder itself. The bottom line is that the risks are much greater than the gains, and ladder pipes should be directed from the ground.

This is not what usually happens on the fireground. When aerial ladders are being used for master streams, firefighters are frequently at the tops of aerial ladders directing the streams. Obviously the decision to go up wasn't well thought out or they wouldn't do it. Why does this happen? Part of the reason is the lack of SOPs or lack of enforcement.

Tradition also plays a role because we have been putting our people up there for years, so it seems like the thing to do. Another reason is that some firefighters *want* to be up there. It seems to them as if they are doing something. They have a great view, and sometimes they can get on the TV news. The downside is that it is unnecessary, dumb and dangerous. Don't take unnecessary chances; direct ladder pipe streams from the ground.

| THIS | NOT THIS |

LADDER PIPE OPERATIONS

Don't Stop the Fire from Venting Itself

When fire is coming through the roof don't use elevated streams to drive the fire back into the building. Just about everybody in the fire service knows not to do this, but it still happens. This incorrect operation stops the natural ventilation of the fire and can cause horizontal fire spread in cockloft and top floor areas. The hole is letting the fire go up and out, which is where we want it to go. The sky is not an exposure. Use ladder pipes or towers to attack the fire through windows or other openings on the fire floor. If most of the roof has burned away, leaving a walled-in bonfire, the fire can then be extinguished from above.

DON'T DRIVE FIRE BACK THROUGH A VENT HOLE

PROPER TOWER LADDER OPERATIONS

Monitor Nozzles

Where master streams are needed in locations where apparatus cannot be positioned, portable deck pipes or monitor nozzles are usually used. Keep the monitor nozzles out of the collapse zone and tie them into a stable object for safety and stability, if possible.

Never Combine Interior and Exterior Attacks

Master streams should never be used along with an interior attack. If you do, people can get hurt, and they will surely be really mad. I once saw an irate Lieutenant confront the IC whom he accused of trying to kill him and his company. It was not a pretty scene. There aren't many hard and fast rules in our business, but this is one is carved in stone. **Never** (as in don't even think about it) **combine attacks.**

Don't Mix Hand Lines with Master Streams

Common sense tells us to use either master streams or hand lines, but in many fire departments, it is a common practice to use pre-connected hand lines along with master streams. One reason for this is that there are usually abandoned hand lines lying around during master stream operations. It takes many firefighters to set up a master stream operation, but once the defensive operation is under way, it doesn't take many people to keep it going. Now you have a bunch of firefighters who want a piece of the action, but there is nothing for them to do.

Too often, Lieutenant Kamikaze and his troops will freelance into the collapse zone using an 1¾ attack line on a fire that Niagara Falls couldn't put out. No gain and lots of potential pain. To keep this from happening, use strict control and SOPs, and outline the collapse zone with yellow tape or cones. Also, use a Safety Officer to enforce the collapse zone.

One exception to the prohibition of mixing master streams and hand lines is when the hand lines are used to protect exposures from a safe position during a master stream operation.

231

Another exception are those rare instances when hand lines can make a difference in the outcome without risks to the firefighters.

DON'T MIX HAND LINES WITH MASTER STREAMS

OFFENSE OR DEFENSE — DON'T KEEP CHANGING

As a company officer, I went to many fires where we were ordered out for a defensive operation. After a little while, we would be ordered back in, only to be quickly driven back out. At some fires, we went through about three or four cycles of changing strategies before it was over. The everybody-in, everybody-out routine is unprofessional and dangerous, and it can cause the troops to lose confidence in your Command. I vowed that if I ever became an IC, I wouldn't play the in-and-out rotation game.

Eventually I became an IC, and when I pull the troops out, I make sure that the fire is controlled before sending them back inside. Making sure isn't that difficult; it just takes water and time. It works because I am aware of the problem and avoid it.

SUMMARY

When it appears that the success of an offensive operation is in doubt, start preparing for master streams.

Procedures for abandoning fire buildings should be simple and understood by everybody.

Before an exterior attack begins, everybody must be out of the fire building.

To set up new defensive sectors, assign companies to the sector where their apparatus is already located, or to the sector that will require the least movement of apparatus.

When preparing for exterior defensive operations, notice the direction the fire is moving and position apparatus and master streams to cut off the fire.

Aerial towers are much more effective than ladder pipes, deck pipes or monitor nozzles because they provide more water and greater mobility.

Direct towers and ladder pipe streams through windows or other openings on the fire floor. Avoid driving the fire back through vent holes on the roof.

Master streams can be very effective for protecting exposures.

Avoid mixing hand lines with master streams, and never combine interior and exterior attacks.

Ladder pipes should be directed from the ground.

17

DISPATCHING AND RESPONSE LEVELS

"A little help at the right time is better than a lot of help at the wrong time." - John Corcoran

Fire departments sometimes fail to dispatch adequate resources to emergency calls. This can seriously affect both firefighter safety and incident outcomes. Most fire department's need to evaluate their dispatch procedures, because too often the procedures are a patchwork quilt of outdated policies from the past.

Dispatch problems are nothing new. In some large fire departments, during the 1930s and 1940s a pulled street box received a response of three engines, two trucks, two battalion chiefs and a division chief. If the fire was phoned in, and no street box was pulled, the alarm response was a single engine, truck and chief. Obviously the dispatching system was screwed up. The reason that this crazy system existed was that the street boxes had preceded the telephone, and in the early years the telegraph was the primary source of alarms. When the telephone came into general use, the dispatch system didn't adjust to the changes.

Many departments today have unrecognized dispatch problems. For example, a single phone call reporting an odor of smoke in a dwelling may receive a standard dispatch of two engines, a truck and a chief. But if the 9-1-1 phone lines light up like a Christmas tree all reporting the same house fire, the response level remains the same in many departments.

There is obviously a problem with this system. It stands to reason that response levels should be increased when there is a strong possibility of a working incident. It is very important to have adequate resources available at our working incidents.

Our first priority is saving lives. Most civilian fire deaths and injuries occur in residences, but many fire departments dispatch heavier assignments to commercial buildings than to house fires. We may dispatch two engines to a house fire where people are reported trapped, while smoke in a 7-11 store receives three engines and a truck company. Obviously this system is out of whack. Part of the reason for this inconsistency is that our dispatch systems are often a spin off of fire insurance ratings that are concerned with protecting against large loss commercial fires. The insurance companies are interested in saving money, while the fire service is into saving people. **Since saving lives is our business, our dispatch policies should value life over property.**

**MOST FIRE DEATHS OCCUR
IN RESIDENCES**
Courtesy of Fairfax County Fire and Rescue,
Fairfax, Virginia

In recent years, there have been many changes in the fire service that require adjustments in dispatch procedures. Some examples include the following:

- Fewer personnel are responding to alarms because of cutbacks in career departments, or reduced membership in volunteer departments, which can result in the need for more companies.

- The Incident Command System (ICS) and NFPA 1500 can increase our personnel requirements. In the past, we seldom used Safety Officers, rapid intervention teams, Logistics Officers, and the other positions needed to manage emergency incidents properly. It is important that fire departments send the resources necessary to cover these important and often mandated positions.

- Hazardous materials incidents are more common and complex, requiring increased response levels.

Too many fire departments have not increased response levels to reflect the changes that have taken place in the fire service.

SEND ENOUGH EQUIPMENT

People can live or die because of the resources available at the scene.

We sometimes don't realize that failing to send enough resources to the scene quickly can result in increased losses. For some reason, many fire service leaders are oblivious to dispatch problems, even when these problems are pointed out to them. It baffles me that the same folks who spend days going over specifications for apparatus, or have prolonged meetings over the color of their trucks are sometimes not very interested in the vital issue of the resources sent to emergencies. People can live or die because of the resources available at the scene.

A common question is: if you send a lot of equipment to the fire, what will happen if there is another fire? The answer is that you already have a fire. Use your resources at the fire you have, not at the fire that you don't have. If there is another fire, there are other companies available, either within the department, or through mutual aid. Adequate coverage should be provided by using an automatic fill-in policy.

Send adequate responses to alarms.

Fireground Resource Requirements

If you cut the resources to the bone, there is no room for error or the unexpected.

The two engines, one truck company and a chief response is probably the most common structure fire response, but it is not adequate for a working fire. The typical house fire is handled with an attack line and a backup line. Additional hose lines may be needed for exposure protection, either internal or external. A rapid intervention team should be standing by for firefighter rescue if that becomes necessary. Truck work at a working house fire usually requires search, vent, laddering, opening concealed spaces, and salvage and utility shut off. Generally two companies will be needed to perform the truck work required within a reasonable time. One chief officer will probably be stressed to the max, so it is a good idea to have two chief officers. In addition, medical support and an air supply truck should be standing by.

I can hear Chief Dinosaur screaming, "That's overkill! We don't have the resources, and we can do it with a lot less." While it may be possible with less, proper resources result in a better and safer operation. If you cut the resources to the bone, there is no room for error or the unexpected. During emergency fireground operations, we must expect the unexpected.

Improving Response Levels Through Flexible Responses

Some progressive fire departments send flexible responses that increase the dispatch when many calls are received, indicating a working incident. One fire department with a total of three engines and one truck company normally sends two engines and the truck to structure fires. When they get multiple phone calls, or when companies arrive on the scene of a working fire, the third engine is automatically dispatched. Fire protection to the town is then provided by mutual aid companies that fill in at the vacant fire stations.

New York City's flexible dispatch policy was used when a rust-bucket freighter floundered just off shore, forcing about three hundred illegal immigrants into the ocean. When the first calls were received by the Borough of Queens Dispatch Center, the dispatchers immediately upgraded the normal first alarm assignment. Instead of two engines, two trucks and a chief, they sent a beefed up response of three engines, two trucks, three heavy rescues, two fireboats and three chiefs. This dispatch judgment and flexibility undoubtedly saved lives, and shows how important it can be to tailor the response to the situation. **The greater the problem, the more you send.**

You don't have to have the biggest fire department in the world to send an adequate response. A small, one station department turns out fifteen or twenty firefighters for a structure fire during evening hours and weekends, but only four or five members during the working week. To ensure adequate responses they have adopted a flexible system that dispatches an additional fire company during low personnel hours. This system provides the personnel necessary for safe and efficient fireground operations.

Good response levels produce good outcomes.

A small career department automatically recalls six or eight off-duty members when they get a report of a working structure fire. They are proactive and don't wait to get behind the curve, and then try to play catch up.

Some departments vary the response depending upon the information received. A report of an appliance fire gets two engines and a truck. Smoke in a dwelling receives three engines, a truck, medic unit and a chief, and reports indicating a working fire get four engines, two trucks, a medic unit and a chief.

DISPATCH ENOUGH EQUIPMENT ON REPORTED WORKING FIRES

Malfunction Junction

In recent years, there has been a large increase in the number of automatic alarms. I was once stationed at a fire station that was known by the troops as "Malfunction Junction" and I became very familiar with the problem. Some fire departments still send heavy responses to automatic alarms, which are almost always accidental or false.

When there is an actual fire, there are almost always additional phone calls. Many departments send one engine company, or an engine and truck company to automatic fire alarms. Some departments have enacted legislation that provides for fines to the owners of alarm systems that have an excessive number of malfunctions. **The less the likelihood of an emergency, the fewer the resources sent.**

Don't Neglect Truck Company Responses

Many departments are weak on truck work; some because they don't have any trucks, others because they don't fully use the trucks that they have. Small departments that lack trucks tend to forget them. They send two engines and a tanker to a house fire, but no truck company, because they don't have one. Their philosophy often is, "If we have a working fire, we'll call for the truck from the next town." If a truck company is available, it should be dispatched initially to structure fires. Where there is not a truck company within a reasonable distance, some of the engine personnel should be assigned to truck functions.

Some departments limit truck responses to buildings of a certain height. For example, buildings over two stories will get a truck response, less that two stories will not. Other departments reserve their trucks for commercial responses, and do not send one to house fires. If the only reason that trucks were dispatched was to be able to use long ladders, the policy makes sense, but that is not the case. Most truck work involves forcible entry, search

238

and rescue, ventilation and all the other things that trucks do. A truck should be dispatched to all structure fires, if possible.

ROLL THE TRUCKS

Trucks in Larger Departments

Many departments dispatch one truck company whether they get one phone call or many calls. I'll bet that your truck staffing is not what you would like it to be. Would dispatching a second truck on calls that appear to be working fires be an advantage? Could another truck company in the early stages make a difference? Could a second truck speed your search operations and ventilation, and improve firefighter safety? If the answer to these questions is yes, perhaps it is time to revise your dispatch policies.

Some departments will provide a second truck company for their mutual aid departments, but fail to send a second truck to their own structure fires. Go figure.

Dispatching Chief Officers

It is important to have an adequate number of chief officers to provide the supervision and control needed at major operations. One department used to send a chief on each alarm and each multiple alarm. Because of a budget crunch, they lost two battalions. As a result, they sent one chief on the first alarm and one on the second alarm, but no more than two battalion chiefs per fire. The idea was to not run out of battalion chiefs. It worked until they had a major interior fire. Command was a disaster, mostly because they had only two battalion chiefs to supervise over 25 fire companies, and one of those chiefs was at the command post. After that fire, the department changed the system back to sending a chief on each alarm.

WORKING FIRE DISPATCHES

It just makes good sense to send resources automatically when there is a working incident.

Many fire departments are now sending a working fire dispatch to provide the additional resources that experience has shown to be needed at a working incident. A typical working fire dispatch includes: air supply, medic unit, an additional chief officer, a Safety Officer, a rapid intervention company, rehab/canteen and a fire investigator. Usually, the more quickly that you get an investigator on the scene, the greater the likelihood that you will find a cause. Besides, you get to go home sooner.

During my early years as a chief officer, I had to remember to ask specifically for each resource needed whenever we had a working fire. It was frustrating to take time from a busy operation to request the obvious at every fire. Eventually we were able to change the system to send a working fire dispatch. This dispatch sends what is needed at a worker in a pre-arranged assignment.

One unnamed large fire department has a problem with sending aerial towers to multiple alarm fires. By the time the IC's remember to call for a tower, it is often too late to get it into position. It was suggested that the response policies be changed so that a tower responded automatically on a second alarm. "No," said the fire chief, "the IC's can call for a tower if they need it." This policy gives the chiefs the freedom to forget to call for towers, and the problem remains to this day.

It just makes good sense to send resources automatically when there is a working incident. Chief Dinosaur says, "I'm the chief, and if I need anything, I'll call for it." That's tough talk, but it often doesn't work very well. Too often, you hear Chief Dinosaur desperately requesting the estimated time of arrival of the air truck, with SCBA bells singing low air in the background.

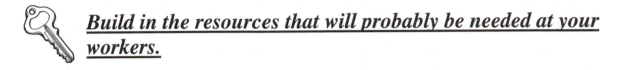

Build in the resources that will probably be needed at your workers.

MUTUAL AID

Citizens in jeopardy deserve the closest available help.

Mutual aid is the answer to many of our resource problems. Some fire departments have a problem with mutual aid. The problem is that sometimes we really don't want **them** in **our** town. (Read as ego and pride.) In the fire service there is no "them," there is just "us." Responses should be based on need and geography, not politics. Ideally, mutual aid should be automatic so as to send the closest stations to respond, even if it is from a different department. Citizens in jeopardy deserve the closest available help. When the fire has you ready to jump out the window, you don't care whose ladder you come down.

It is essential to have mutual aid agreements in place to be able to use all the resources available. These agreements should include which fire companies respond on each alarm. This system will also avoid fire companies self-dispatching to the scene of major emergencies. Self-dispatchers are well intended, but the result is added confusion to an already difficult situation. Clear mutual aid agreements should state that help is not provided unless called for.

DISPATCHING MULTIPLE ALARMS

Prepare for the worst and send adequate help to multiple alarms.

Greater alarm responses should generally be the same as or more than the original response. This approach should provide sufficient personnel and reserve to properly handle a working incident. For example, a standard second alarm might bring three engines, two truck companies, a chief officer and a medical unit. This response allows the IC to summon a predetermined amount of resources with a single command. This avoids the "nickel and dime" approach of dribbling in help, or the "Send me everything you can get," message that a desperate IC requested after a general alarm was sounded.

1ST ALARM	ENGINES 11, 21, 9	TRUCKS 6, 9	B.L.S. 2	CHIEF 4
2ND ALARM	ENGINES 4, 1, 16	TRUCKS 14, 11	A.L.S 33	CHIEF 6
3RD ALARM	ENGINES 14, 24, 6	TRUCKS 4, 2	B.L.S. 9	CHIEF 1

SAMPLE ALARM ASSIGNMENTS

I subscribe to a fire newsletter that lists major multiple alarm fires occurring in their region. The dispatch to second alarms varies so much that it is hard to believe that the fire departments are all serving similar towns in the same area. A second alarm response may consist of a single engine company, an engine and truck, or two engines and a truck, or various combinations up to four engines and two trucks.

Fires don't really care about town boundaries. If a worker requires four engines, two trucks and two chief officers, it doesn't matter what town the fire is located. Since the resource requirements are similar, why do response levels vary so greatly? The answer is usually found in the size of the fire departments. Small departments typically send smaller responses even though they are part of a big network. The reasons are probably a mix of tradition and their philosophy that they can handle it themselves; no outsiders are going to mess with their fires if they can possibly help it. **There is no such thing as a small fire department. Some just think small.**

IF IT'S NOT A FIRE, WE OFTEN UNDER-DISPATCH

People in peril should get a response that is sufficient to handle their problems.

A child entered a large drainage pipe and walked a considerable distance before falling into a 12-foot pit. Dispatch handled the incident as an injured person from a fall and dispatched an ambulance. Much more help was needed for the rescue.

An auto ran into a house, breaking a natural gas line in the house, and spilling gasoline inside the house. In addition, there were injured people needing treatment. The Dispatch Center sent an engine and medic to this incident. Clearly the response was not adequate.

A call to the 9-1-1 Dispatch Center reported a strong odor of gas in the area. Two police cars were dispatched to investigate. After their arrival, an explosion occurred and the police officers requested that the fire department respond. That scenario made about as much sense as sending a fire company to investigate the sound of gunshots.

More People Die in Car Wrecks than Fires

Lives often hang in the balance at vehicle accidents.

Many more people die in auto accidents (around 44,000 deaths annually) than fires (about 4,300), and our dispatch policies should reflect this fact. We usually send more equipment to fire calls than to vehicle accidents. Think about it. An odor of smoke in a house often gets two or three engines, a truck and a chief officer. A head-on collision often gets a response of an engine company, a medical unit and, sometimes, a heavy rescue. The everyday fender bender may not require a large response, but serious accidents (which are normally obvious from the number and nature of the phone calls received) should receive an adequate response.

For vehicle accident responses, don't settle for what you've always done, or what you can get away with. Instead consider the best way to operate. The ICS system should be used to manage accident scenes. Typical sectors include: Medical, Extrication, Safety and Suppression. These are not textbook sectors. The chances are that you will need all of them to handle a serious accident effectively. Think about it. Which one do you think you can let slide because it's not really that important? Each of these sectors increase efficiency, improve the safety of both victims and rescuers, and allows the IC to maintain an overview of operations. The problem is that staffing these sectors requires more resources than are typically at the scene of most accidents.

SERIOUS ACCIDENTS REQUIRE MANY RESOURCES

It is common in some departments for chief officers to be dispatched on automatic fire alarms, but not to vehicle accidents with reports of persons trapped. The automatic alarm is very likely false or accidental, and seldom needs a command officer. A serious vehicle accident often requires a strong command structure for scene safety, triage, extrication, transportation, helicopter operations and many other tasks.

In addition, a chief officer should respond to serious vehicle accidents because the chief is going to be in charge when a mass casualty occurs. If they don't command the smaller incidents (where they gain valuable experience), it is not reasonable to expect chief officers to manage major incidents properly. A good rule of thumb is to have a chief officer respond to reports of major vehicle accidents.

DISPATCH CHIEFS TO SERIOUS ACCIDENTS

Lives often hang in the balance at vehicle accidents, so we should send sufficient resources to handle the reported emergency. The response to serious accidents should generally be equal to your department's structure fire response, with the addition of a medic unit and a heavy rescue squad.

Haz Mat Responses

Haz mat incidents are often under-dispatched. For example, the fire department of a medium-sized city received a 9-1-1 call reporting an unknown chemical spill in a building. The fire dispatch sent a single engine company to investigate. If the call was for smoke, they would have assumed that it was a fire and sent a full response. It is lunacy to apply different standards to haz mat, and play "lets go see if it's really a problem." There is no way that a single engine can properly handle a haz mat incident. In this case, there was a working haz mat incident, and the department had to play catch up. This approach reminds me of an old policy when the police were sent to investigate the need for an ambulance. If the police found that Uncle Charley was really having a heart attack, they would call for an ambulance. Under this system, Uncle Charley often died by the time the ambulance got there.

Fire departments should send sufficient resources to handle the reported emergency. Many fire departments send a standard initial response to investigate any suspected or potential hazardous materials incident. A typical response consists of a chief, engine, truck, ambulance, and the Hazardous Materials Team. This response level recognizes the seriousness of hazardous materials calls. Reports of larger incidents receive increased response levels.

PRIORITIES IN CALLING FOR THE POLICE AND UTILITIES

Sometimes the need for a rapid response is critical.

Many incidents require a police or a utility company response. Most requests are routine, such as needing the police for traffic control, or the power company to shut down a meter. Occasionally, however, the need for a rapid response is critical. You may need the police to respond quickly because you are being assaulted, or to assist you with an evacuation. There may be a critical need for the power company because electrical arcing is going to result in the loss of a building if it is not quickly controlled.

Many departments lack a priority system to ensure quick responses. Sometimes departments think that they have a priority system but they actually don't. For example, one fire department uses the term "expedite" when they need the police to respond quickly. The problem is that "expedite" is not a written policy understood by all, so it may or may not cause a rapid police response.

The best way to handle these problems is to establish a priority system that is understood by all. The term "Priority 1 response" can be used when there is a critical need for the police or utility company to respond quickly. For example: *We need the police Priority 1 for an evacuation.*

MOVE-UPS AND FILLING IN VACANT STATIONS

When working incidents have companies committed for long periods, other companies are often assigned to fill in at the vacant stations to provide fire protection to the area involved.

Most departments have delegated the fill-in responsibilities to the Dispatch Center. In some departments, however, the responsibility for move-ups falls on the incident commander of the working incident. The IC is very busy trying to keep on top of the incident, and does not have the time to worry about empty fire stations.

I have had numerous discussions with people from the "chief makes the fill-ins" departments, but I have never heard a convincing argument for why it is done. Usually the argument goes, "Well, we've always done it," or "The IC is the only one who knows how long companies will be on the scene." This system is probably from the days when small towns did not have Dispatch Centers, and telephone operators turned out the fire trucks. A phone operator was not qualified to make fill-ins, but a modern Dispatch Center certainly is. To have the Dispatch Center make the fill-ins simply requires establishing guidelines for the dispatchers. We need to do everything we can to make the job of the IC easier.

DISPATCH PROBLEMS

Most fire departments have some problems with the dispatch process. When you talk with fire personnel about dispatching, you usually hear much griping. If you talk to dispatchers, they often aren't too happy with us firefighters. Most of the problems between dispatchers and firefighters result from a lack of communication (no pun intended.) A few departments are fortunate enough to have some dispatchers who are fire buffs at heart, but most departments are working with police-type dispatchers. These dispatchers often have little understanding of fire department operations.

Dispatchers often lack firm guidelines on what questions to ask callers. It is important to ask key questions to determine the situation and proper response levels. Dispatchers must relay all pertinent information to the responding companies. This includes reports of people trapped or injured and the location of the fire. There have been major lawsuits over the deaths of people who notified the Dispatch Center that they were trapped at fires, but the information was never relayed to the fire fighting forces.

Engine 1 was dispatched for an automatic alarm in an office building. Dispatch then called the engine that was still en route and said: *We now have a telephone call reporting a fire in the building, do you want us to dispatch a full assignment?* This question shows the lack of good procedures. Response levels shouldn't be up to whoever is sitting in the officer seat of Engine 1 that day. The additional phone call should be handled by dispatch procedures. With good SOPs in place, the message should have been: *Dispatch to Engine 1, we now have a report of a fire in the building and we are dispatching a full first alarm assignment.*

I was responding to an alarm when the radio dispatcher told me that they were receiving multiple calls, and they believed it to be a working fire (it was.) Upon return to quarters, the dispatcher called me to ask if that information had helped me. I told him that it had because I set up a staging area in advance and was better prepared. The dispatcher had gotten into trouble for taking that initiative because they didn't know for sure that it was a working fire. When a Dispatch Center lacks a written policy, expect inconsistencies in the dispatch process.

Dispatchers should have adequate training and clear dispatch standards, and they should be able to change or upgrade assignments based on the circumstances.

How to Improve Fire Dispatch

Some departments have improved their dispatching by forming a steering committee (including dispatchers) to review existing guidelines and to make recommendations. The best guidelines are usually simple. Don't get too fancy. Trying to fine tune the dispatch process so that each type of call gets a different response can result in a dozen or more different response patterns that are confusing to both dispatchers and responding firefighters. Simple is better. For example, structural fires, reports of serious accidents or hazardous materials incidents, special operations (technical rescue) and water rescues should receive a similar response. All of these incidents require many resources and equipment to operate safely and efficiently. Think about this. Which of these incidents is so simple that you need few resources?

SIMPLIFY DISPATCH PROCEDURES
Courtesy of Fairfax County, Fire and Rescue,
Fairfax, Virginia

Provide ride-along programs for dispatchers so they can observe our operations and our culture (or, sometimes, the lack of it.) Initiate tours of the Dispatch Center for fire personnel, so that we can gain appreciation for their demanding jobs. Schedule periodic meetings between fire and dispatch personnel to improve operations. Fire dispatchers should feel that they are a part of the fire department team.

SUMMARY

Send adequate responses to alarms.

Response levels should be increased when there is a strong possibility of a working incident, or when there are shortages of personnel or water.

Most civilian fire deaths and injuries occur in residential occupancies, not commercial buildings, and response levels to residences should reflect this.

Update your dispatch process to utilize truck companies fully.

Greater alarm responses should generally be the same or more than the original response.

Reports of serious accidents, hazardous materials incidents, special operations (technical rescue) and water rescues should receive a response similar to your structure fire responses.

Almost all fire departments are part of a big dispatch network.

Establish and maintain mutual aid agreements.

Periodically review and update the dispatch process.

245

18

HAZARDOUS MATERIALS

The first rule of haz mat is similar to the Hippocratic oath, "First, do no harm."

Many firefighters find it hard to understand why any other firefighters would want to deal with pesticides, yellow smoke, and radiation, when fire fighting is where it's really at. In some fire departments, our haz mat personnel are known by derogatory terms such as "the glow worms." This seems to reflect our bewilderment and uncertainty about this complex subject. While there are entire books on hazardous material response, some are long on theory and short on reality. Most of us can use all the help that we can get when it comes to dealing with haz mat operations. This chapter covers the basics of haz mat operations learned from working at hazardous materials incidents.

Pre-Planning Haz Mat

Avoid having to play catch up.

"You can't pre-plan haz mat incidents," says Chief Dinosaur. "They are all different." The Chief is off-base because most hazardous materials incidents follow predictable patterns. Almost half of the Washington, D.C. Fire Department Hazardous Materials Company's calls involve gasoline, diesel, or propane fuels. Other common responses are for releases of chlorine, ammonia, acetylene, and hydrochloric acid. Your department's experiences are probably similar and reasonably predictable.

Most of your incidents will involve hazardous materials made, stored, or shipped in your area. Locations and frequencies of your previous incidents will give you a good idea of the types of incidents you will come across in the future. Review shipping routes, industrial occupancies, or illegal dumping areas. You can then focus your training, pre-plans, and preparation on your most likely scenarios. This will allow you to be proactive instead of reactive. As the saying goes, nine-tenths of wisdom is being wise on time.

Get All the Information You Can

The more information you have, the better off you are.

Before or while responding, the IC should talk to Dispatch to get more information. Dispatch usually knows a lot more than they could tell you at the time of dispatch. The more information that you have, the better off you are.

One afternoon, we were dispatched for a "spill" with no other information. Further inquiry enroute revealed that it was reported to be a mercury spill in a laboratory. After consulting with the responding haz mat company by radio, I notified the first due engine company that the spill was mercury, and directed them to stay out of the building if there was no life hazard. It turned out that there were no lives at risk, so the first and only company to go inside was the fully protected haz mat company. Advance knowledge was used to reduce the risks to our personnel.

First on the Scene

Haz mat companies are almost never the first to arrive; fire companies are. The actions of the first company often affect incident outcomes. Because their initial actions are critical, all companies need to know the basics of haz mat operations.

Positioning is important. Arriving companies have been known to drive right through spills. The first arriving company should respond from upwind, isolate the area, identify the product, and report the situation. This size-up report can help all the responding companies by giving them a jump on the situation. No company should enter the hazard area except for rescue. Remember that fire fighting gear is not designed to protect the wearer from hazardous materials.

Every vehicle must have a current copy of the Emergency Response Guidebook (developed by federal transportation officials) — for a free copy call your state coordinator listed in http://hazmat.dot.gov/gydebook.htm

Look for placards on transportation vehicles and look for NFPA diamonds on structures or fences. If the incident involves a transportation vehicle, the driver should have shipping papers listing the contents of the load and material safety data sheets for the materials that are defined as hazardous. At a fixed location, material safety data sheets should be available. Of course, drivers or knowledgeable on-site personnel may be absent, dead, or otherwise unable to give you information. Gather and analyze as much information as you can before you act.

LOOK FOR NFPA DIAMONDS ON STRUCTURES

Companies other than the first due engine should be dispatched to a staging area, not to the actual incident. This simple procedure can keep most of our people out of harm's way.

STAGING
AREA

THE FIRST ENGINE GOES TO THE INCIDENT

**DISPATCH COMPANIES OTHER THAN THE FIRST DUE
ENGINE TO A HAZ MAT STAGING AREA.**

The Sneaky Haz Mat

A cardinal rule is to recognize hazardous materials incidents.

When we think of haz mat, we often picture a large plume of green smoke coming from a tank car on its side. In the real world, most incidents are a lot less dramatic. Most incidents are not big and obvious, and, if you're not careful, they can sneak up on you.

One afternoon, I was the IC at an incident where there was an unknown odor in an office building. The odor was from a stopped-up drain that the maintenance people were trying to clear with chemicals. Nobody was excited or even very concerned; a ho-hum attitude prevailed. The maintenance people were clearly annoyed that we were there, and the office workers seemed nonchalant because the odor was not very strong. The fire companies were treating it more as a plumbing problem than a haz mat incident.

My guard was down, and little by little, we started to get in deeper and deeper. Reading the label on the can of what they poured down the drain showed that the fumes were poisonous. A call to Chemtrec (Chemical Transportation Emergency Center) revealed that we had more of a problem than we thought. Meanwhile office workers continued to walk past the drain room. About this time, I finally realized that we were in the middle of a haz mat incident and started to play catch up, doing many things that I should have done from the beginning.

At another incident, a lawn care truck left a pesticide spill in the road. I considered closing down the street, but that seemed unnecessary. It was just a small spill near the curb line on a residential street with little traffic. While we were looking up the product information, a car entered the block and made a U-turn right through the spill! **The moral is: Take nothing for granted.**

Beware of hazardous materials incidents that masquerade as fires. One morning, we had a fire involving some propane cylinders. We had a haz mat incident, but most of the troops thought we had a fire. Too often when the troops see fire, they think fire, and when they see skull and crossbones or purple smoke, they think haz mat. This shouldn't happen, but it does. We need to improve our awareness and safety at fire-related hazardous materials incidents.

Think worst case scenario.

A cardinal rule of haz mat is to recognize the potential severity of hazardous materials incidents. Two important haz mat principles are: (1) take it seriously, and (2) set up monitored control zones and make them larger than you think you need. **It is better to begin seriously and later discover that the problem is minimal, than it is to screw around until you begin to realize that you are in trouble.**

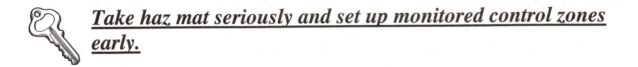

Take haz mat seriously and set up monitored control zones early.

First Of All, Do No Harm

All hazardous materials incidents eventually self-terminate.

The first rule of haz mat is similar to the Hippocratic oath, "First, do no harm." This means that sometimes the best decision is to take no action, or limited action. Estimate what would happen without fire department intervention. Remember that all hazardous material incidents eventually self-terminate.

If intervention is necessary, risk as little as possible. Just because a hazardous material is on fire doesn't necessarily mean that we should put it out. Some fires in flammable metals and chemicals may be better if left to burn themselves out. When water is necessary, consider unmanned master streams because they are safer than hand lines.

Another cardinal rule at haz mat incidents is to use the minimum personnel necessary in the hazard zone. Firefighters putting up the yellow barrier tape often believe that the tape is for civilians, not for them. This may be true in some situations, but at haz mat incidents, the yellow tape is for everyone. We need to educate our people that control zones are for their protection.

USE MINIMAL PERSONNEL IN THE HOT ZONE
Courtesy of Fairfax County Fire and Rescue, Fairfax, Virginia

Estimate harm if no action is taken. If you must act, use the minimum personnel necessary in any hazard zone.

The Mystery of Chemistry

Chemistry is almost never used at actual incidents.

Many responders feel inadequate at hazardous materials incidents because they do not know or remember the chemistry of hazardous materials. Chemistry is seldom used at actual incidents. You don't need to know the principles of internal combustion to drive your car. All the technical information you need is in the haz mat reference books. Besides, you can't work from memory, even if you think you know the product. The best approach to handling hazardous materials is not chemistry; it's doing the basics and using the reference materials.

Develop the Haz Mat Plan

Get information from as many sources as possible.

Your plan should start with a good SOP for handling hazardous materials incidents. Consider your previous experiences as a guide, but don't assume that similar incidents will be the same as the last one. To assist you in developing your plan, first identify the product and the quantity involved. Get information from as many sources as possible, keeping in mind that you should not rely totally on any one source. Check with:

- The occupants or people involved because they often know what happened and may be knowledgeable about the product

- Labels, but remember that the manufacturer may tell you only what they want you to know

- Placards (use binoculars if necessary)

- Haz mat guide books

- Chemtrec (800-424-9300)

Your command chart can be helpful in the development of your plan because it should show:

- A brief description of the product involved

- A diagram of the area

- A check-off list of haz mat reminders

Have the haz mat officer recommend a plan for handling the hazardous material, and get input from your key players. The IC makes the final decisions, but it is important to discuss all the opinions and information so you can make the best decisions. If something doesn't seem to make sense, it may not. Be skeptical and ask questions. Some good questions are "Why?" "What other options are there?" and "What will happen if we don't do anything?" If you don't understand something ask to have it explained again so that you understand. This is much better than pretending that you understand. No one expects you to be a haz mat expert.

Determine the worst case scenario and plan for it. For example, if a gasoline tanker is on its side with minimal leakage from the dome covers, the worst case is a release of all the gasoline, which will result in either a fire or flowing gasoline. Don't just plan for what you have. Plan for what you could have, and you will seldom be caught short.

A basic sample plan for a major gasoline spill in the street follows:

THE PLAN

- **PREPARE FOR IGNITION**

- **EVACUATE AREA**

- **ESTABLISH CONTROL ZONES**

- **PROVIDE FOAM BLANKET**

- **SAFETY FIRST**

- **PROTECT EXPOSURES**

- **PROTECT ENVIRONMENT**

HAZARDOUS MATERIALS INCIDENTS MUST BE WELL PLANNED

It is important to always have a backup plan, a lesson that I learned the hard way. The incident involved crystallized ether, which can explode like a bomb. We had a great plan to evacuate the building involved and to protect the bomb squad, but we had no plan about what we would do if it had exploded. All our plans were aimed at safe removal, not the possibility of an explosion.

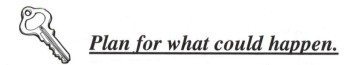 ## _Plan for what could happen._

A written plan is the best way to cover everything. Gather all the players and explain the plan, and what their roles are. Make sure that everybody is singing the same song. Meet periodically with haz mat personnel and the Sector Officers to discuss the plan and the progress being made (or not made.)

Control Zones

The control zones should be set up early and at the proper distances.

Time after time, the lack of control zones causes problems at hazardous materials incidents. The zones should be set up early and at the proper distances. Set up physical boundaries to visually indicate the zones, using haz mat barrier tape, cones, lifelines, or other barriers to clearly show the hot, cold, and warm zones. The hot zone is the hazard area and the warm zone is where decontamination can take place. Only in the cold zone can personnel operate without protective equipment. The hot and warm zones should have only one entry and exit point, because control is difficult if there are numerous access points. Once the zones are established, guard them with constant vigilance. No one should enter the hot zone except the hazardous materials personnel. It may be necessary to use fire companies to secure the area until sufficient police are present.

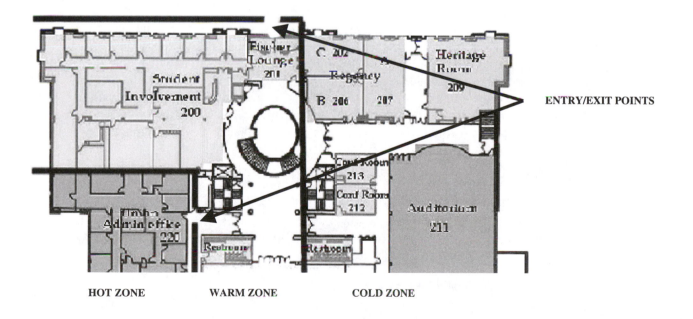

HOT ZONE   WARM ZONE   COLD ZONE

AN EXAMPLE OF CONTROL ZONES INSIDE A BUILDING

Evacuations

Begin the evacuation with those who are most threatened. This appears to be a no brainer, but I recall one evacuation of a city block that started at the end of the block and worked its way toward the problem area.

The evacuation distances recommended in the haz mat guide books are often not practical in urban areas. For example, if the recommended distance is 2,000 yards, there is no way that you can evacuate all the people within that distance in a city environment. Often the best that you can do is to evacuate those in imminent danger, and shelter the rest in place. Protecting in place requires notifying civilians, instructing them to close windows and doors and shut off ventilation systems that use outside air, and updating them with new information. Police officers can assist you with this work.

EVACUATIONS ARE DIFFICULT IN A CITY ENVIRONMENT

One night, there was a working fire in a commercial building near a large hospital. The fire involved some hazardous materials and smoke was drifting towards the hospital. The decision was made to protect-in-place, because it was safer to leave the patients in their rooms than to move them. The outside air intakes to the hospital were closed off; the outside doors facing the smoke were kept closed, and the inside air was monitored. It worked. If this had been a private house, it may have been safer and easier to evacuate the residents, but evacuation of the hospital would place more people in danger than if they had remained in their rooms.

Evacuation Problems

There always seems to be a back door at every incident.

One afternoon, we had a fire involving propane cylinders and other hazardous materials. The incident was outdoors, next to a large college dormitory in the middle of a busy campus. We needed to cool the cylinders, evacuate the dorm building, and keep bystanders away from the incident.

In the early stages, we didn't have enough people do everything that we needed to do (sound familiar?) We used the campus police to keep back the bystanders and to evacuate the dorm. It didn't work very well because the private security force was your basic rent-a-cop outfit, and left a lot to be desired. Some of the students left the dorm, but others didn't. Some who

254

stayed would periodically lean out their windows overlooking the burning cylinders. We did the best that we could with what we had, and it turned out that we were able to stabilize the propane cylinders before we could complete the evacuation.

Evacuation is difficult, and it is often not possible to know for sure that everyone is out. I remember one haz mat incident when a civilian wandered out of a supposedly evacuated apartment building in the middle of the operation. He had been asleep in his apartment. It was not practical to force open every door in the apartment building to guarantee complete evacuation.

It is one thing to evacuate and another to keep it evacuated. Large areas or buildings with many entrances are hard to police over long periods of time. There always seems to be a back door at every incident. It is important to maintain a close liaison with the commanding police officer on the scene, and to check periodically that the security is still in place and maintained. **The moral is: Be prepared for difficult and complex evacuations, and consider protecting in place.**

Evacuate Once

At some incidents, the same people have been evacuated two or three times.

Determine where the people being evacuated will be relocated to and how you will get them there. Often that means using buses to a school, church, or public building. It is important to evacuate people just once. There should be no doubt that the evacuation center that you have chosen is far enough removed that it won't have to be re-evacuated. At some incidents, the same people have been evacuated two or three times because the IC failed to estimate the potential of the incident. Think of the logistical nightmare of repeated evacuations of the same people. It is hard enough to evacuate once, let alone two or three times. **Do it right, and evacuate once.**

Commanding Haz Mat Incidents

Commanding haz mat incidents can be difficult. Most chief officers haven't commanded enough hazardous materials incidents to maintain the skills needed. The following common sense suggestions can help to make the IC's job a little easier:

- Make sure that control zones are established and maintained

- When there is the possibility of fire, have fire protection ready for immediate use

- Verification of the product and its quantity is very important. Verify, verify, and verify. At one incident, the product had a label, so we read what the label said about the product. We then double checked with Chemtrec and then triple checked with lab personnel on the scene. All three sources concurred, so we were reasonably sure of what we were dealing with

- Chemtrec is an invaluable source and often provides information that is not in your hazardous materials library. They also give emergency medical assistance advice from experts in the field. Chemtrec may be useful, even if you think that you know all the answers. At one incident, Chemtrec advised us to upgrade our entry suits to a higher level than our references recommended. At another incident Chemtrec concurred with the private contractor to mix the product with dirt and then remove it. It would have been unwise to rely totally on either the reference book or the contractor

- Require the haz mat company to complete a hazardous materials properties sheet for the command post. This sheet shows the properties of the product(s) that you are dealing with. This is important. Without it, you're flying blind. It is a good idea to keep somebody from the haz mat team at the command post if possible, to keep the IC informed

- Assign a team to scout the area to make sure that the incident is confined. Hazardous materials can spread through the air and by liquid runoff (sometimes through sewers.) Checking around keeps you ahead of the curve and avoids nasty surprises. If the incident covers a wide geographical area, consider using a helicopter for observation

- Look over the situation and come up with a broad strategic plan of what you are trying to accomplish, and have a backup plan just in case

- Use a laid back team approach and take your time. Consider everything, check, double check, and triple check

- Make a work list of those things that are completed, those still being worked on, and those remaining to be done. Make another list showing the questions that still need to be addressed

- Haz mat incidents get very busy, and it can be an effort to keep priorities straight. At one incident, we had firefighters with burning sensations on the skin who were in danger of being put on the back burner because of the volume of incoming information and the number of people involved. It may be necessary to call a time out to focus on the key issues

- Haz mat incidents often require a great deal of communications, so try to use face-to-face communications whenever possible. Spell out the names of products. Many chemicals have similar names even though the properties may be very different

- Use a Safety Officer and always weigh the risks against the gain

- Make sure you have adequate medical capability on scene

- Additional chief officers are often important for safe and effective operations

- An adequate reserve should be on hand, and a fully prepared backup team should visually monitor those in the hot zone

- Consider decontamination procedures early to avoid problems later in the operation

- At a working haz mat incident, the IC can quickly become overwhelmed by the media, outside agencies, and the police. The IC can relieve a lot of pressure by establishing both Liaison and Information Officers to handle these concerns. Limiting access to the command post also helps

- Haz mat incidents produce much stress and uncertainty. Everything seems to take longer than it should. Don't rush haz mat; we are aiming for safety not speed. Time is usually an ally. Some of our troops just want to flush it and go home. It is difficult to avoid pressure from the media, the troops, and politicians to hurry up and get the job done. These folks may mean well, but I don't think that they would be willing to rush their own open heart surgery. Safe and efficient operations are the sole responsibility of the IC. Don't let anyone talk you out of doing the right thing, because if the thing blows up, you will be all alone

- Don't let the Feds, police, or others take stupid chances. Keep an eye on them

- In some places, the police department or other agency may be the lead agency for handling haz mat incidents. When this happens, the fire department will need to use diplomacy to work with the lead agency

- Avoid the tendency to relax and let your guard down as time goes on. Haz mat incidents have a way of dragging on, especially if you handle them properly. Your plan should also consider the logistics of food, toilets, and rehab

Don't rush haz mat. Time is usually an ally.

Who Is In Charge?

The IC is a generalist, and the haz mat officer is a specialist. The average IC does not have the haz mat training that the haz mat officer has. The result is that the IC is often hesitant to change, modify, or over rule the recommendation of the haz mat officer.

The haz mat officer knows that the IC is in charge. The haz mat officer may also assume that the IC won't permit unwise or unsafe operations, though this is not necessarily true. There can be a dangerous situation when the two principal players each think that the other knows best.

I had an incident when the presence of a knowledgeable haz mat officer intimidated me into agreeing to a shaky operation. The officer recommended wearing low level protective clothing. It seemed to me that it would be wise to move up one level to be on the safe side, but I went along with him. After the incident was over, I realized that safety was compromised and that I should have overruled the officer and required use of a higher level of protection. There was no disadvantage to using a higher level, and we would have provided better protection for the haz mat crew.

Following this incident, I learned my lesson. Later, at another incident, a haz mat officer said that he did not recommend using SCBA because the product was not toxic, but could be an irritant under some circumstances. I heard him out and concluded that there was no disadvantage to using SCBA, which would increase our safety margin if something went wrong. I made the decision to use the best protection available. After the incident, the haz mat officer said, "I hate to admit it, but you were right about the SCBA protection we used." The moral is that haz mat people are skilled, but they are not responsible for the safety and welfare of everybody at the scene. The IC must not delegate the running of the incident to the haz mat officer by approving whatever haz mat recommends.

Listening to the Experts

If something doesn't seem to make sense, it may not.

A qualified hazardous materials expert can be invaluable. Try to develop working relationships in advance with haz mat experts who can help you when you need them. The IC should consult with experts, but must remember that the ultimate responsibility remains with the IC.

When you must deal with unknown experts, a few basic questions can help. Ask the expert about their prior experiences. Do a common sense field test by asking the unknown expert a few questions about the incident that you already know the answers to. These questions can help you get a feel for the credibility and reliability of the expert.

Commanding haz mat incidents is difficult, so it is easy to hope that somebody will come along to help resolve the situation. Some years ago we had a fire in some drums that had been abandoned by a midnight dumper. We were unsure of the contents so an "expert" chemist arrived to help us. He walked into the hot zone stuck his finger in the drum and licked it. He told us, "It doesn't taste toxic." So much for unknown experts.

The experts can easily become part of the problem instead of the solution. There are lots of people out there who are incompetent or have their own agendas. The expert's agenda may be to reduce liability for whomever the expert represents, or to save money for his organization. Some people just want to feel important, and they will take over if they can. Some people from outside agencies will turn a fire department operation into a circus if you let them.

Document the Incident

Documentation is important. Assign someone to keep written notes of command post activity for an on scene reference. A tape recorder, camcorder, and still photographs can also be useful. All important information, such as entry and exit times, should be recorded. This can be valuable if medical problems arise at the incident or in the future. Documentation can also be used during the post incident analysis, to help improve future operations.

DOCUMENT HAZARDOUS MATERIALS INCIDENTS

Documentation can sometimes be used for reimbursement from the people who caused the problem. Many fire departments routinely bill businesses to recover the costs of providing haz mat services.

Disposal: Who Says It's our Job?

Cleanup belongs with those who made the mess.

The fire department's job is to mitigate emergencies. Once the haz mat scene has been stabilized, it is no longer an emergency. Avoid moving or disposing of hazardous materials. The cleanup job belongs to the owner, occupant, shipper, or cleanup contractor.

SUMMARY

Pre-plan the types of hazardous materials incidents likely to occur in your area.

Get all the information you can before arrival.

Companies other than the first due should be dispatched to a staging area, not to the actual incident.

Take all haz mat incidents seriously. Upon arrival, isolate the area, identify the product, determine the quantity involved and evaluate the situation.

Verify the information you receive. Chemtrec is an invaluable source that often provides information that is not in your hazardous materials library.

Set up big control zones and monitor them. Use haz mat barrier tape, cones, or other barriers that clearly show the hot, cold and warm zones.

Estimate harm if no action is taken. If you must take action, slow and safe is the best way.

Use the minimum number of personnel necessary.

A basic haz mat principle is to determine the worst case scenario and plan for it.

Consider protecting in place.

Evacuate early and once.

Assign a scout team to check the area to make sure that the incident is confined. Incidents can spread through the air and liquid runoff.

Document the incident.

Avoid the tendency to relax and let your guard down as time goes on.

19

SPECIAL OPERATIONS

"No one tests the depth of a river with both feet." - Ashanti proverb.

Special operations (or technical rescue) includes water rescue, high angle rescue, and hostile situations. All are dangerous and highly specialized operations. Each of these could be covered in a book of its own, so this chapter covers only the basics of some of the most common special operations. However, the concepts and philosophy of all special operations are similar.

In all special operations, the first priority is the life safety of the rescuers. This may be difficult for us to accept, but we simply can't go jumping into rivers or off buildings in "shoot-from-the-hip" rescue attempts.

Organize Special Operations Teams Properly

In most cases, a regional team is the most practical and cost effective.

Special operations teams are highly specialized and require major commitments of time, personnel and money to maintain the top notch teams necessary for these incidents. Too often these teams are started with more enthusiasm than forethought. It is easy to get sucked into the glamour of this rescue work, but the hard realities must be faced. Most departments do not have enough special operations calls to keep up the skills of the team personnel, or to be cost effective. The reality is that in most cases, a regional team is the most practical.

Get All the Information You Can

Before you go diving into a river in the middle of the night, you want to make very sure that the car hasn't been under water for a week.

While responding, question Dispatch to find out all the information available about the incident. Upon arrival, look for and question witnesses to find out what occurred, when it happened, and the status of possible victims. Because you see one person in the river doesn't necessarily mean that there is only one victim. Because there is a car in the pond, it doesn't necessarily mean that it's occupied. Special operations are dangerous enough, so before you go diving into a river in the middle of the night, you want to make very sure that the car hasn't been under water for a week.

Commanding Special Operations

Upon arrival at the scene, the IC should consider:

- What information do we have?

- What are we trying to accomplish?

- Are the risks worth the gain?

Special operations must be managed by using both ICS and SOPs. When it is necessary to attempt a rescue, use the minimum number of personnel in hazardous positions. Use a backup rescue team at all special operations, as well as a Safety Officer.

Don't lock into one method of rescue. Have a backup plan. For example, if you are planning to use a boat for a water rescue, consider calling for a helicopter so you don't have all your eggs in one basket. Special operations often require large numbers of personnel and resources for correct and safe management, so make sure that you have an adequate response to handle the situation. During long duration operations, rotate the rescue teams.

The chief officer is not expected to know everything about these operations. The IC should trust the expertise of the specialized teams (they know what they are doing) but apply situational judgment (is this the right place to do it?) The IC needs to listen to the troops and avoid high risk operations when the troops are uncomfortable with them. On the other hand, the IC may need to calm down some of the kamikaze types who are looking for an opportunity to do their thing. Too often specialized teams will take unnecessary chances to justify their teams. They may want to go diving into a river at 3 a.m. for a body recovery because they have been in training for years. They think it will be good experience, and besides, they may not have another opportunity for a long time.

In all special operations, the first priority is the life safety of the rescuers.

WATER RESCUE

An Invitation to Die.

"I can still see the faces of my firefighters as they were swept down the river." (Lieutenant Milton A. Dofflemyer, Rescue 2, District of Columbia Fire Department.) These two firefighters were attempting the rapid water rescue of people trapped on top of a car in a river. The rescue boat capsized, spilling the two rescuers into the river. Fortunately, both firefighters were able to cling to debris and tree branches and were rescued.

In another incident, a vehicle was submerged in a flooded creek, with a second hand report of a person still in the truck. An experienced fire department diver went into the water, but did not use a backup. He never came up and died in the tangled wreckage. The vehicle was abandoned and had been in the water for several days. The diver had skipped all the common sense precautions and simply jumped in.

Firefighter SCUBA team members sometimes die while performing body recoveries. In one case, a firefighter died trying to recover the body of a drowning victim in an abandoned quarry. Recently an empty canoe was found on a river, its owner presumed drowned. For several days, fire department SCUBA teams dove randomly into the river since the location of the body was unknown. The risks of this "needle in the haystack" diving were great, with nothing to be gained. In about five days, the body surfaced on its own.

THE LAST RESORT IS TO GO IN THE WATER
Courtesy of Cindy Beasley, RQ3 (rescue3va.com)

Some years ago when I was a brand new captain at the heavy rescue, we had a rescue boat that was covered with dust. Not knowing much about boating, much less emergency boat rescues, I asked an old salt rescue lieutenant about boat rescue work. He said: "My advice is that if you get a white water rescue call don't take the boat." I laughed thinking that he was kidding. He was serious. His reasoning was we were in a city environment, had a boat that wasn't designed for rescue work and had little training. After listening to him I concluded that he was probably right. I started the process of getting a suitable boat, water rescue SOPs, and proper training.

The cardinal rule of water rescue is **don't go in the water if the rescue can be made any other way.** The last resort is to go in the water. Unfortunately, the untrained often consider this as the first option. Use the low risk options first.

Low risk options include trying to reach victims with poles, ladders, or lines. In some circumstances boats or helicopters may be the best options for a safe rescue operation. When it is necessary for firefighters to go in the water, use every precaution, know what you are doing and why. For example, a safety rope on the rescuer may be appropriate for a pond rescue, but is dangerous for rapid water rescues. **Water rescues are very dangerous. Use your head not your emotions.**

Adequate Responses

There are few situations with more dangers to victims and firefighters than water rescues; don't downplay the danger.

Water rescues usually require lots of personnel and equipment. For a river rescue, you will probably need personnel on the opposite shore (it may be closer), downstream (victims could be carried there), upstream (for spotting floating debris headed toward the rescue site), and for backup (for the rescuers.) A truck company should be at water rescues because the ladders (or bucket) can be a lifesaver. A helicopter, a boat, and medical units are often needed. All of this probably sounds like overkill if you are not familiar with the hazards of water rescues, but there are few situations with more dangers to victims and firefighters, and we should not downplay them. Many water rescues involve resources from several jurisdictions so it is important to have SOPs and mutual aid radio channels in place. These procedures should be reinforced by multi-jurisdictional drills.

Be Prepared

Most places have rivers, streams, or dams that have the potential for water rescues. Most water rescue incidents are predictable because they follow patterns. The locations of the waterways in your area as well as your past experiences will give you a good idea of the type of incidents that you will probably come across. You can then focus your training, pre-plans, and preparations on your most likely scenarios.

Fire departments faced with water-related emergencies should develop SOPs and provide equipment and training to accomplish the rescues that are common in their areas. Most departments, however, have infrequent water rescues, and have little water rescue equipment or training to use it. Articles in our trade journals that show you how to blow up a section of hose with a SCBA bottle for a water rescue show how ill prepared we can be.

Don't do what one fire department did. A river in their city had a vicious drowning machine, a low-head dam with a hydraulic action from which it was almost impossible to escape. Predictably, one day they had a rafter caught up in the hydraulic. They arrived with little equipment, no training, and no plan. Later one firefighter said, "It was a terribly helpless feeling, knowing that somebody is going to die because you can't get him off the river." They stood on the shore, not knowing how to handle this very dangerous situation. They mulled over a few options and drew a blank. After about ten minutes of head scratching, some TV news personnel in a helicopter heard the situation on their scanner, called the fire department and asked if their helicopter could be of any assistance. Yes, yes, yes was the reply. The coptor arrived and in five minutes; the victim was safely on the shore. The person almost died from the fire department's lack of planning.

Prepare the Fire Companies First

Fire companies arrive first at nearly all water rescues.

To accomplish water rescues, we need to have the proper equipment and the expertise to use it. Usually we put most of our resources and training into a specialized water rescue team, but fire companies arrive first at nearly all water rescues. These first arriving firefighters often have little training or equipment. They sometimes get caught up in the situation and, without thinking, just jump in the water. It is very important to use a life jacket (and a helmet for swift water) when entering the water. All fire companies need to be prepared with training, equipment, and SOPs. "Whenever we have heavy rains and the potential for flooding" said one firefighter, "they send me to the shed to find the life jackets and rope. We have no SOPs or training. We just might need them." Unfortunately, this firefighter could have been talking about many fire departments.

At one incident I witnessed, two people were trapped on top of a car in a rapid water situation. A fire officer tied on a lifeline and jumped in the water slightly *downstream* of the car. The "rescuer" somehow made it to the car and climbed on the roof. The next arriving company now had three victims to rescue instead of two. This situation was caused by the lack of experience, training, and equipment, all of which is a recipe for disaster. Obviously we cannot turn every company into a SCUBA team, but we can prepare them for the dangerous water rescues that they may face.

HIGH ANGLE ROPE RESCUE

High angle rope rescue should be the last resort.

Too often fire department high angle teams will rappel down the front of a building to rescue stranded window washers. It looks impressive on the TV news, but it is usually not necessary. Normally, mechanical adjustments to the lift can be made to rescue the washers. When this doesn't work, consideration should be given to bringing the window washers in through the windows. This is seldom done because of the fear of excessive damage to the windows, as well as the danger of falling glass. On the other hand, we seem willing to have our people (and the window washers) on the end of a rope hanging in the air. If we can teach people to safely jump off the top of buildings, we should be able to train them to safely remove high-rise glass. The high-angle rope rescue should be the last resort, never the first. **The moral is to think about ways to avoid high-angle rope rescues.**

The Saga of the Cliff Woman

When a fire department establishes a high-angle rescue team, they recruit volunteers to staff the team. The folks who volunteer are going to be macho and aggressive. (The timid don't do high-wire acts.) These teams spend much time practicing, but seldom get a real life situation. The result is that all too often these folks are looking for excuses to do their high-wire act.

Not long ago, a woman told her family that she was going to go for a walk. When she failed to return, a major search was undertaken. Shortly before she disappeared, she had been in a park with cliffs overlooking a river. The police believed that she possibly jumped or was thrown off the cliffs to the river and rocks below. The fire department high-angle team was called to the scene. The police chief wanted the team to scale the cliffs to look at the bottom, which could not be seen from the top, and most of the team was all for it. Here was an opportunity to show off their skills. One member disagreed, saying that the evidence was flimsy and the cliff dangerous. Besides even if she was there, it would be a body recovery operation, not a rescue. He suggested that a boat be used to look below the cliff. Fortunately the cooler head prevailed, and the team did not go over the side. The boat confirmed that there was no body. Several days later the missing lady returned. It seems that she had a lot on her mind and had been off meditating someplace.

DON'T HANG FROM A ROPE IF YOU CAN AVOID IT

FIRE DEPARTMENT OPERATIONS AT HOSTILE INCIDENTS

We often force our way into buildings for lockouts, lock-ins, or other dubious reasons. Whenever we force our way into a home, anyone inside may consider us an intruder, shoot first and ask questions later. Often, we shouldn't be using forcible entry in the first place, so thoroughly question the occupant to see if the need is legitimate.

If it is necessary to go inside, take some basic precautions. Always have the police on the scene before sending people in. There could be a domestic problem, or the person may not belong there. Find out if there are any occupants inside, awake or asleep. Ask if there are any guns or dogs inside. You get the picture. Don't go in if you can avoid it. If you must go inside, be very cautious. If you are called to a lock-out under false pretenses, and there is no unattended baby or food on the stove, don't let the occupant in. The last thing we need is people who endanger us by lying and breaking the law, and getting away with it. The false caller should be turned over to the police for legal action.

**BE SURE THE SCENE
IS SAFE**

Hostile incidents also include barricade or hostage situations, bomb threats and terrorist activities. Unfortunately, these calls are becoming more common. These situations are police matters, but we can get involved in them by providing medical support, and by being ready to assist in case of a fire or explosion. A chief officer should respond to these types of hostile incidents for coordination, safety, and control.

Another area of concern is response to violent medical calls. Calls for shootings, stabbings, and other assaults can be dangerous to responding firefighters. Many progressive departments are issuing bulletproof vests to both fire and medical personnel.

When responding to calls reported to be violent, companies should stage a block away until police have arrived **and** declared the scene as secure. There have been incidents where firefighters were shot because they assumed that, if the police were on the scene, the scene must be secure. **WRONG!** Assume nothing. In these situations the safety of our firefighters is paramount, and the key to safety is good SOPs, tight control, and staging.

Don't Let Your Firefighters Become Cowboys

Firefighters should be out of the action until the situation is secured.

A firefighter was assisting the police by climbing a ladder to enter a building where a deranged person was holed up. The firefighter was shot dead. Police actions are not like the movies. People bleed and die, and there is much grief for everyone. Our firefighters must be out of the action until the situation is secured. Firefighters should not be climbing ladders in hostile situations. I'm not sure that the police should be climbing ladders either. When we are sent to help the police with a ladder, we tend to go everywhere our equipment does, and we can get shot as a result. Firefighters must be out of shooting range.

If He Comes At Us with a Gun, Open the Nozzle

The call was for a barricade situation, with a lone gunman holed up in a house. The engine company was dispatched directly to the scene, without staging or a chief officer. Upon arrival, there was a fire in the dwelling and an armed gunman was loose somewhere in the house. The police SWAT team was outfitted with SCBA, and the engine company went in with their attack line, with police with drawn guns on either side of the pipe man. The fire lieutenant told the firefighter, "If he comes at us with the gun, hit him in the chest with a straight stream." They went in, and found that the gunman had used the gun on himself and was dead. Later, in discussing the incident, most of the firefighters thought that because the fire was extinguished without injuries, that it was a good job. **Think danger when police SWAT teams are used.**

HOSTILE INCIDENTS ARE VERY DANGEROUS
Courtesy of the Columbus, Indiana, Police Department

Chancy Evacuation

Both the firefighters and the residents would have been a hell of a lot safer if they had stayed in bed.

The police SWAT team was going into a three-story apartment building to capture some very dangerous people. The fire department was called to the scene to assist them. The police wanted the building empty of all the people in the building, except for the apartment with the bad guys. They had the gunmen isolated, and they intended storming in with guns and tear gas. There was little likelihood that the gunman could get into the hallway, let alone the other apartments. The IC nevertheless agreed to evacuate the building, so firefighters put up ladders and knocked on windows in the middle of the night to wake up the residents (also a good way to get shot.) They then escorted the residents down ladders to the ground. Unbelievably, this action was assumed to be safe. Both the firefighters and the residents would have been a hell of a lot safer if they had stayed in bed.

Police Operations

Most police are simply not used to operating at large complex operations.

The police sometimes handle large scale hostile situations poorly. The tragic incidents at Waco and Ruby Ridge attest to that. I have seen many occasions when the police have been unorganized and wandering around in the potential line of fire. The only reason that the police were not shot was that the gunman didn't see them or didn't want to shoot them; it was certainly not lack of opportunity.

One reason that the police are sometimes unorganized is the nature of their business. Most police actions are after the fact. The bad guys don't often hang around waiting for the police. They are often long gone before the police arrive. Another reason is that most police actions only involve a few officers at each incident. Hostile situations usually involve large numbers of police, who are often not used to operating at large complex operations, as the following scenario shows. Currently many police departments are improving their operations by using police ICS and developing SOPs for handling incidents, so we should be seeing improvements.

A Car Full of Dynamite

The police were scared.

The car was reported to be loaded with dynamite, and an armed loony was in the car, ready to blow it up for some obscure reason. The police were scared and believed that the guy was serious. They had reason to believe that he really had the dynamite. A standoff continued for many hours, when suddenly there was a rapid burst of gunfire from the SWAT team. It took the firefighters *and the other police* completely by surprise. Several firefighters were in the line of fire, but fortunately they were not hit. They were on their way to supper, had become careless, and were driving through a restricted area.

The police command post soon noticed what they believed to be smoke rising from the car. They told the fire department IC that the car was on fire, and they wanted it extinguished. The IC asked two questions: 'What is the status of the dynamite?" and 'What is the status of the gunman?" They didn't know the answer to either question, but wanted the fire department to move in. The IC replied that the fire department was not going to take any action until they were absolutely certain that the scene was safe. This story has a happy ending (except for the guy in the car who was killed.) It turned out there was no dynamite and no fire, just steam from stray shots that had hit the radiator.

The moral is don't accept anything that the police or anyone else tells you at face value. Ask questions and verify everything. If the police say that the area is secure and they want us to go in for medical assistance or some other reason, it is wise to be skeptical. Ask questions: 'Why is it safe?" 'Is the suspect in custody?" 'How do you know that there aren't any more suspects?" You get the idea. Verify, and then verify again. We can't entirely trust the police or anybody else, and it is our responsibility to protect our own people.

Hostile incidents are serious business. They literally hiss with danger, and we need to be very careful. Don't let a macho attitude get in the way of safety. Strict control must be exercised. No freelancing can be allowed. Rule number one in your SOPs should be to use a staging area for control and to stay in a sheltered position. This puts everyone, including the police, on notice that we are not placing our people at risk. Let the police handle the police work. When these incidents involve a standoff of an hour or more, consider staging the fire companies at the nearest fire station. The IC and a fire officer can remain with the police.

 ### *At hostile incidents, use staging areas and stay in sheltered positions.*

Communications are vital, but all too often at these incidents, we aren't sure what is happening. Maintain a constant liaison with the police by assigning a fire officer to the police command post and a police officer with a police radio at the fire command post. Better communications and improved safety are the result. It is important to keep the troops informed of the situation and the game plan.

During civil disturbances, firefighters should be under police protection, and all companies should respond as a group or task force under a chief officer. Under these conditions, we would only respond to structure fires. Task force commanders have the authority to pull out of, or not enter into, any situation when the risks are too great.

Terrorism

Many of these incidents can be handled using current procedures. For example, the ICS system can be used on any type of incident. Fires are handled using SOPs, and biological agents and radiological incidents are essentially hazardous material incidents. There can be multiple problems such as mass casualties associated with a hazardous materials incident. In those cases, we need to handle the multiple problems with a single integrated plan.

Large scale terrorism incidents require many resources, including many more mutual aid companies than you are used to. This usually results in coordination and communications problems. A unified command is essential, as is a logistics officer. Think about things like how and where you can get a dozen backhoes and two giant construction cranes on short notice. In addition, communication with law enforcement, hospitals, public health officials, and the media is essential.

Being prepared also means thinking about what has happened and what could happen and figuring out how you can best handle the incident if it occurred. The most likely places for terrorism are prominent places that have the potential for a lot of casualties and damage. Accountability can be a nightmare, so practice it with multiple jurisdictions and agencies. Large scale disaster drills are the order of the day, and every emergency responder should be trained in basic terrorism awareness.

Terrorism Response Considerations

- Establish staging areas at a safe distance, and allow no freelancing

- Take time to perform an adequate size-up. Quickly evaluate the severity and magnitude of the incident. Question people on the scene to determine what happened prior to arrival (vapor clouds, suspicious people, unusual odors, the presence of chemical devices, etc.)

- Always consider the safety of emergency response personnel, weighing the risks against the benefits. Use minimum personnel in high risk areas, and have RITs standing by

- Personnel accountability is essential, use accountability officers

- Evaluate structural stability

- Establish collapse and control zones early

- Work with law enforcement to deny entry into damaged buildings

- Assign a team to observe ongoing activities in and around the incident to observe the potential for a secondary event

- Consider using apparatus PA systems or bullhorns to communicate with people during evacuations, triage, decon etc.

- Company officers must do what they are assigned, even if it's not the glory job that will be on CNN

- Use police to assist in crowd and traffic control

269

- Consider activation of mass casualty and disaster plans

- Use personnel experienced in detection and monitoring equipment to check for chemical, biological or radiological releases as needed

Fully utilize the incident command system. Typical branches or sectors required are the following:

       Staging
       Evacuation
       Suppression
       Search and Rescue
       Safety
       Medical
       Hazmat and Decon
       Liaison

SUMMARY

Develop guidelines and SOPs for handling all special operations, but recognize that it is impossible to set a rigid framework of procedures for every situation.

Train on the basics of special operations.

Get as much information as you can about what occurred, when it happened, and the status of possible victims.

Put safety first, exercise a strong command, use common sense, and maintain good communications.

Always weigh the risks against the benefits. Always use the minimum number of personnel in the hazard area.

The IC must manage special operations incidents using ICS and SOPs.

Don't lock into one method of rescue. Always consider alternate means.

Conduct reconnaissance at all special operations.

Use a backup rescue team and a Safety Officer at special operation incidents.

20

DE-ESCALATION OF THE INCIDENT

'It ain't over till it's over."- Yogi Berra

Demobilization is often a weak link in fire department operations. Too often we fold the command and safety tents after the incident is under control. Yet, many firefighters are killed or injured after incidents are "controlled." Since planning and SOPs are important parts of well-run incidents, it simply makes good sense to wind the incident down the same way—according to a plan. The philosophy of the chief should be: let's take it slow and easy, and make sure that no one gets hurt. Keep the Safety Officer and RIT in place until safety is assured. Secure or cordon off unsafe areas and keep enough personnel on the scene to do what is needed.

A Typical Relief Problem

The time of dispatch for a working fire in a commercial building was 20:02 hours and involved four engines and two truck companies. A canteen truck was requested to respond at 22:01 hours (almost two hours later.) The last company left at 01:52 hours (almost seven hours later) There was no relief, and the troops were beat to death. This situation is typical of the relief problems in some fire departments. Fire companies should be relieved regularly for safety, efficiency, and morale.

Setting Up Rehab

The first commandment for a fireground commander is: Take care of your troops.

One evening, several firefighters were talking around the kitchen table. "You know," said one 'have you ever noticed when we have a B.S. room fire we are rehabbed to death. Every time you turn around someone wants to take your temperature." Another firefighter chimed in, "Yes, but when we had the nasty fire at the hardware store , there was no rehab and we were there all night."

Why does this happen? A major reason why rehab can be slow to non-existent is because there are not enough people on the scene to allow relief. Another reason is that many fire departments are programmed to rehab after extinguishment, regardless of what the SOPs say.

No Extinguishment, No Rehab.

Rotating crews into rehab requires an adequate reserve, especially when the incident is still not under control. Using the ICS system, the IC should have an adequate reserve and a Rehab Officer, otherwise rehab probably won't happen. The Rehab Officer maintains a schedule based on the time a company has been on scene and the type of duty performed. Companies pulling ceilings may need rehab after thirty minutes, while those protecting an exposure may be able to wait an hour or more.

A Rehabilitation Sector should be established at every working incident because exhausted firefighters are more likely to get hurt. When your department uses a system where the elapsed time at an incident is announced regularly (as recommended in the Dispatch chapter), the need for relief is less likely to sneak up on Command.

Rehab should be set up far enough away from the incident so that personnel can remove protective clothing. The rehab area should be clearly marked with a flag or barrier tape. Consider using vehicles or buses for shelter during extreme weather. One department says that they can't call for an empty bus because they will be charged for it. The answer is to either pay for it or make other arrangements. It is unacceptable to have our people freezing or baking on the streets because the fire department brass have not provided for the welfare of their people.

Companies should stay together in Rehab because they may unexpectedly be called back into action. Rehab should last for at least fifteen or twenty minutes, longer if necessary.

 ## Set up rehab at all your workers.

In Rehab, paramedics should give medical checks to the troops, and any abnormal signs or injuries should be treated. An air supply truck should be near rehab to refill SCBAs, and a canteen should provide food and drink. **Rehab shows the troops that the brass are concerned about their well-being and safety, improving morale as well as safety.**

TAKE CARE OF YOUR TROOPS

First In, First Out

Exhausted companies should be returned as soon as possible.

Too often, the first companies to arrive are the last to go home. This leaves the first arriving troops, who are wiped out, to handle a big overhaul and salvage job. This policy doesn't make much sense, but it's a common practice in many departments. When they finally do get back to the fire station, if another fire or emergency occurs, the exhausted firefighters are not going to be much help.

Tired firefighters are more likely to be injured if they are used for the overhaul. The best way to determine who needs to be returned is to ask the Sector Officers. On smaller jobs you can ask the companies themselves, but be aware that few companies will admit needing relief. An officer sitting on the front lawn looking more like a victim than a firefighter, will usually say that his company is okay.

FIRST IN ➡️   ⬅️ **FIRST OUT**

Sometimes, an IC will continue to use exhausted troops because their apparatus is surrounded with other apparatus and difficult to get out. The health and safety of our people should come ahead of the logistical problems of apparatus placement. There are several ways to return surrounded companies quickly. One way is to have them switch apparatus with a later arriving company who can get out easily. Another way is to transport exhausted companies back to the station, leaving their apparatus and equipment. A fresh company can be assigned in their place to complete the overhaul.

The SCBA One Hour Bottle Rule

After companies use SCBA for one hour, they should be returned to the station if possible.

"It was a really tough fire, and so I used four SCBA bottles," commented one firefighter after the fire. Superman wouldn't be super on his third and fourth bottles. At working fires, it is not unusual for firefighters to use multiple air bottles and be totally exhausted. At the same time, there may be other firefighters not called, wishing they were at the fire. This is poor management of our most precious resources, our people.

Crews should go into rehab after one 30-minute air bottle is used, (or its equivalent in a one hour bottle.) In addition, there should be a one hour bottle **limit** for each company. This means that when a company has used two 30-minute bottles or one 60-minute bottle, they should be returned to the station if possible. "That's not prac tical,"says Chief Dinosaur. "The place is burning down, and my resources are limited."There may be times when this is true, but what it usually means is that the IC has not called for help early and in force.

The Two Hour Rule

After two hours, the troops often are tired, careless, and unenthusiastic.

After two hours of fireground work, the average crew is exhausted. A good rule of thumb is to use a fire company for no more than two hours, and considerably less when they have engaged in heavy work. When the troops are tired, they often become careless, unenthusiastic and prone to injury. Few have heard of the two hour rule because I made it up after years of experience and observation. I use the two hour rule, and it works. There is a strong need for some sort of standard. We have all sorts of personnel and tactical standards, but none to protect our troops from total exhaustion. The two hour rule is particularly necessary when the weather is bad. You often hear about considering the weather in size-up, but you seldom hear about it being a consideration in the relief of companies. Always consider the effect of weather on your troops.

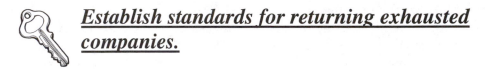

Establish standards for returning exhausted companies.

Return Companies Slowly and Wisely

We had just returned from a multiple alarm fire. The lone engine company still on the scene called for a full first alarm assignment to come back to assist them. Off we went, back to the fire we had just left. Going over the road, I couldn't imagine what was going on. The fire building was just about a total burnout, and the fire appeared to be dead cold out. As we got closer, I couldn't believe my eyes. The sky was lit from the glow. I was totally baffled. Upon our arrival, the mystery ended. The fire wasn't in the original building. It was in a large old wooden water tank that was on top of the exposure building. The tank was higher than the fire building, and apparently it had been heated to the smoldering point. Just as the last company was about to leave, poof! It had erupted into a ball of flame. Fortunately we extinguished it with little damage to the building.

Don't be too eager to return fire companies. Sometimes unexpected things happen. Check all the angles first, such as smoke conditions, exposures, and the amount of overhaul needed. Let the dust settle for a while. If you return companies too quickly, you may find yourself in the embarrassing position of having to ask for more companies after you have declared control and returned companies.

Canteen Blues

I recently talked with a volunteer who works on the canteen of a big city fire department. He said that whether the canteen is called or not depends more on who the chief is than what the incident is. Often, he and the other volunteers will hear incidents on the radio that sound as if they may need the canteen. The volunteers call the Dispatch Center to have them ask the IC if the canteen is needed. The canteen gets about half of their runs this way, indicating a rehab problem in this department.

Canteen responses should be part of the dispatch protocol. The canteen should roll on all working fires and multiple alarms automatically. It should also be used on long duration incidents such as brush fires, and when the troops are exposed to extreme weather conditions.

SUMMARY

An organized demobilization provides for the safety and welfare of our personnel and the orderly return of unneeded resources.

A Rehab Sector should be established at every working incident.

Rehab should have shelter, a medical unit, an air supply truck, and a canteen to provide food and drink.

Companies should go to rehab after using SCBA for 30 minutes. When a company has used SCBA air for a total of one hour, they should be returned to the station if possible.

A good rule of thumb is to use a fire company for no more than two hours, and considerably less when they have engaged in heavy work.

Don't be too quick to return fire companies. Unexpected things sometimes happen.

Make sure that the first companies to arrive at working incidents are the first to return to their stations.

21

POST INCIDENT ANALYSIS

'Don't find fault. Find a remedy."- Henry Ford

Military training uses the classic battles to teach their commanders. Studying past battles is the key to understanding future battles. The fire service should use the same principles and conduct post incident analysis to improve the effectiveness of our operations. We are condemned to repeat our past mistakes if we don't learn from them. We should find out what went wrong to keep it from happening again, and find out what went right to make it happen again.

How do we know how well managed our firegrounds are? The answer is that we usually don't really know. People have different opinions, and we often lack any real system to evaluate our emergency operations. Too often when things go badly on the fireground, we fail to find out what went wrong. Our post incident analyses (PIAs) are often weak to non-existent.

We have all kinds of graphs and charts showing us where firefighter deaths are occurring, but little information showing us the strategic and tactical errors that are responsible for these deaths. I have worked on firefighter death probes, and have studied the United States Fire Administration Technical Reports investigating firefighter fatalities. Three of the most common factors contributing to firefighters deaths are:

- Poor communications

- Lack of accountability

- Poor strategy and tactics

All three are command functions, indicating problems in command systems. These problems also indicate the lack of both good SOPs and post incident analysis systems. Too often, we see the same problems repeated over and over. We need to improve our operations in order to increase the odds that everyone will make it back to the station after every call.

The Close Calls Keep Coming

A good evaluation system should be the early warning system that something is wrong.

A large fire department had a fire where a severe backdraft occurred, resulting in firefighter fatalities. Later, it came to light that similar backdrafts had occurred previously in other areas of the city, but were not reported because nobody had been injured. Question: Why didn't the department use PIAs to recognize and correct the problems before somebody got killed? Firefighters seldom die from problems that are new to the department. When poor communication leads to the death of a firefighter, you can bet the farm that communications are a problem in the day to day activities of that department.

Several years ago, a fire department had a fire in an abandoned, boarded-up building. To attack the fire, they made a rat-hole entrance, a very small hole in the side of the building, just big enough for the attack crew to crawl through. No other openings were made, and there was no ventilation or other exits. The inside of the building was a maze and so smoky that the crew was blind. The fire flashed over, driving the crew from the building. One firefighter became disoriented and barely made it out of the building. It was a very close call. There was no PIA, so the safety problems went unrecognized, and no operational changes were made.

Several years later, it was deja vu all over again. The same department had a very similar fire, and they did the very same thing, sending an attack crew into a rat-hole opening. This time the fire didn't flash over, but the potential was there for the same thing to happen again, with the possible loss of the engine crew. The lack of a PIA of the first fire could have resulted in injuries or death at the second fire. The second fire was not evaluated either, so the potential for disaster still remains in this department. Too often, somebody has to die before we start looking critically at our operations.

 ### Don't allow uncorrected problems to cause firefighter deaths.

Avoiding the PIA

Poorly run departments don't learn from their own mistakes.

Why do many fire service leaders shy away from the PIA? The answer is often fear. The biggest fears are that they will look stupid or that the firehouse lawyer types will cause dissent and trouble. There is also the fear that they may wind up with a final report that makes them or their fire departments look bad.

One unnamed fire department had a major fire that was captured on video. The operation was lousy. After the fire was controlled and they had returned to the station, one of the firefighters asked if there would be a PIA of the fire, especially since they had access to the fire video. The officer replied that the chief was saying that they might, but that he was trying to talk the chief out of it. They never did hold a post incident analysis. Sharp fire departments learn from the mistakes of others, average departments learn from their own mistakes, poorly run departments don't learn from their own mistakes.

In another department, following a major structural collapse resulting in a mass casualty incident, a deputy chief was quoted as saying, 'Everything went great. There was nothing that we could have done better.''It is unusual for everything t o be letter perfect at any incident, and impossible at a major one.

Following a major fire in a department that has very few big fires, the fire companies were required to complete a PIA for the apparatus repair shop. The shop was an outside agency that had responded to the fire to maintain the equipment, and was not part of the fire department. The repair shop wanted to learn from the incident to improve their shop operations. The fire was never reviewed by the fire department. What effect do you think this had on the morale of the troops?

Some common excuses why a PIA is not held:

- There was nothing we could do. (The fires control us.)

- These things happen. (We have to expect lousy operations.)

- There was too much fire. (We can only learn from small fires.)

- Everything went fine. (You can't improve on perfection.)

- We don't have time to do a PIA. (Emergency operations are low priority.)

THE POST INCIDENT ANALYSIS

Every incident is an opportunity to learn.

Holding PIAs routinely is a key to success. Every working incident should be critiqued. When the PIA is held routinely, it is no longer a big event where no one is quite sure what will happen. By doing PIAs regularly, the troops won't be thinking, 'Something went wrong and they are looking to place blame.''

Regular PIAs can reduce freelancing because Lieutenant Kamikaze soon learns that what he does at an incident will become part of the record. The regular PIA also keeps the incident commanders aware of what's happening on their firegrounds. Without the PIA, it's easy for IC to be unaware of potential problem areas.

The ICS system can be used as a basis for evaluation. Study what parts of the command system were used, what wasn't, and why. A review of the departments SOPs can be used to compare what happened with what was supposed to happen. If something was not SOP but worked well, or if the SOP operation was poor, it may be time to review the SOP to see if it needs to be changed or updated. The idea is to reinforce good techniques and change those that are not working.

It is important to review incidents quickly. If you let too much time go by, memories fade and stories change. The purpose of the PIA should not be to head-hunt or sweep things under the carpet, but to improve your operations. While the post incident analysis can show problem areas, they can also show what went right. It is important to praise and reinforce what went right.

Hold PIAs routinely to improve your operations.

Post Incident Analysis Methods

There are several different methods of conducting PIAs, including formal analysis, informal reviews, written evaluations, and interviews.

Formal Post Incident Analysis Meetings

Prepare in advance for the PIA. Dispatch tapes, fireground channel audiotapes, pictures, diagrams, and video footage can be a part of the learning process. When conducting a post incident analysis, it is important to have a relaxed non-threatening atmosphere, where people can open up and be candid. Everybody should get their say, but some control may be necessary so the meeting is not taken over by those with the biggest mouths. If that happens, the PIA can become a finger-pointing, shouting match, with more heat than light, resulting in dissension and a muddle of what really happened.

A FORMAL POST INCIDENT ANALYSIS MEETING
Courtesy of Montgomery County Fire Department,
Montgomery County, Maryland

When reviewing incidents, the IC can set the tone by relating the lessons that command learned from the incident. This openness expressed by the IC will show the troops that this is going to be a worthwhile meeting and not the usual bureaucratic "it wasn't my fault shuffle." Safety should be the number one priority when reviewing an incident, so the Safety Officer should be present. Focusing on safety will also reassure the troops because everyone considers safety a high priority item that should be addressed. The PIA leader should ask questions to generate a discussion, such as "Lieutenant Rowan, looking back at your ventilation effort, is there anything that you could have done to improve your operation?" At the conclusion of the PIA, the leader should summarize the incident, including what was done well and what needs improvement. Following the meeting, a written summary of the incident and the lessons learned should be published and distributed. **The moral is: A good post incident analysis will improve operations, safety, and morale.**

Informal Post Incident Analysis

Individual firefighters can learn their own lessons by thinking about past incidents and by studying fire magazines and critiques from other fire departments. Listening to fire radios, watching video tapes of fires, and discussing incidents with other firefighters also helps. Informal critiques are a good way to involve the front line firefighters. Company officers should review even basic incidents within their companies. An engine company back from an automobile fire can talk about what happened. Sure, car fires are usually simple, but we tend to get careless at our routine operations. Every incident is an opportunity to learn. Play the 'what if' game following the PIA. If the fire involved one room and was quickly knocked down, discuss what you would have done if the fire had extended.

AN INFORMAL PIA

EVEN A CAR FIRE CAN BE A LEARNING EXPERIENCE
Courtesy of Alan Etter DC Fire/EMS

Every incident is an opportunity to learn.

Written Evaluations

The dullest pencil is better than the sharpest mind when it comes to retention.

I first learned the value of a written evaluation when I was assigned to downtown Washington's 6th Battalion. The chiefs on the other shifts were progressive and eager to learn, so after each fire I wrote up what happened and the lessons learned. Although my intent was to assist the other chiefs with what I had learned, I soon found that I was the biggest beneficiary. By writing and following a standard format, I was forcing myself to consider every phase of the operation in depth.

Written evaluations are usually more thorough than oral ones, and can remain on record for future use. The dullest pencil is better than the sharpest mind when it comes to retention. Firefighters who were not at the PIA may not learn anything unless the results are written. In large cities, there can be a serious fire with firefighters injuries and most members won't even know about it unless there is a written evaluation. In most departments, if you were not at the fire, you are often clueless about the lessons learned. The written evaluation also allows other fire departments to learn from your experiences, if you are willing to share with them. Finally, a written evaluation can be an effective rumor control, to set the record straight on what happened.

**WRITTEN EVALUATIONS
IMPROVE LEARNING**

One heads-up firefighter gripes, 'Today's technology allows us to communicate with other fire companies and jurisdictions by using computers, e-mail, faxes, and so on."But he went on to say, 'Not once have I ever seen any lessons learned at incidents com e across any line of communication available to us today."Technology sharing is a great asset in the private sector, but the fire service is years behind. We need to catch up.

Comparison Analysis

The comparison analysis is a relatively new post incident analysis system. Following each working incident, reports are submitted by all the company officers, Sector Officers, sub-commanders (Operations, Logistics, Safety, etc.) and the Incident Commander. All officers report what they did, why they did it, problems found, and results. Company officers should complete their reports with input from their personnel.

When all the forms are completed, they are compared. If the forms are similar with few problems, it can be assumed that the incident was handled effectively. When there are some differences between the reports or problems are brought up, further investigation is needed. Major discrepancies between reports indicate major problems.

The written report is objective, reduces hot tempers, and forces each officer to think about all aspects of the operation. Comparative analysis is often an easy way to begin using a post incident analysis system.

 **PIAs save lives.**

The IC's Report Card

To stay in business, you must know what is happening and what your strengths and weaknesses are.

The post incident report card can help the IC to evaluate incidents. After each working incident, the IC completes the report card. The card considers all areas of fireground operation and standardizes the evaluation process. Each function, such as fireground ventilation, is evaluated by assigning it a rating from A to D, with A being best. In time, you will have enough data to compare your incidents and look for patterns. This will allow you to see both strengths and weaknesses of your operation. These ratings clearly show where you fall short and improvements are needed. This is a systems approach. In completing the report card, the IC must be objective and consistent. Ego doesn't belong in an unbiased evaluation. It may help objectivity if the IC does not share the card, but keeps it for personal improvement.

 Incident commanders must use a system to evaluate their performance.

I used the following report card to rate the efficiency of firegrounds at incidents where I was the IC. The following card shows fires soon after I began using the rating system, and there was plenty of room for improvement. As you can see, the weakest areas (highlighted) were: Protecting exposures, testing the air, the IC position, and salvage. On the basis of this information, I focused on the weak areas to improve them. Total scores were obtained by assigning values: A=4, B=3, C=2 and D=1.

FIRE LOCATION	1718 P St. 6-story apartment	3200 O St. 3-story townhouse	1700 M St 10-story office bldg	4304 12 St 2-story house	1840 2nd St. 2-story townhouse
IC POSITION	(D)	(D)	C	C	C
SOP FOLLOWED	C	B	A	B	B
REAR COVERED	C	C	B	C	B
ROOF REPORT	B	C	C	B	C
EXPOSURES PROTECTED	(D)	(D)	C	C	(D)
SEARCHES COMPLETE	B	B	A	B	B
MEDICAL COVERED	B	B	B	C	C
USED A PLAN	C	C	B	B	B
BACKUP PLAN	(D)	(D)	B	C	C
SAFETY FIRST	C	C	B	B	C
AIR TEST	(D)	(D)	C	(D)	B
UTILITY CONTROL	C	(D)	C	B	C
SALVAGE	(D)	(D)	C	B	B
REHAB & RETURNING	B	B	C	C	B
INVESTIGATE CAUSE	B	B	B	B	(D)
TOTAL SCORE	29	29	40	37	35

The following shows a report card from one year later with significant improvements in fireground operations. These improvements were a direct result of using a standard criteria to evaluate fireground operations. Notice that the total scores improved considerably over the previous year. Also notice that the improved total scores are very similar, with little variation. Consistency is the hallmark of a properly functioning system.

FIRE LOCATION	6634 Georgia 2-story apt house	1334 Quincy 2 ½-story townhouse	5237 Iowa 2-story house	1350 Clifton 5-story apt house	1705 Lanier 6-story apt house
IC POSITION	C	B	B	B	B
SOP FOLLOWED	B	A	B	C	B
REAR COVERED	B	B	B	B	B
ROOF REPORT	A	A	A	B	C
EXPOSURES PROTECTED	B	B	B	B	B
SEARCHES COMPLETE	B	B	B	B	B
MEDICAL COVERED	B	B	B	A	B
USED A PLAN	B	B	B	A	C
BACKUP PLAN	C	C	B	C	C
SAFETY FIRST	A	B	B	B	B
AIR TEST	B	B	B	C	B
UTILITY CONTROL	B	B	B	B	B
SALVAGE	B	B	B	B	B
REHAB & RETURNING	B	B	B	B	B
INVESTIGATE CAUSE	B	B	B	B	D
TOTAL SCORE	45	46	46	44	42

Similar systems are used every day in the business world. To stay in business, you must know what is happening and your strengths and weaknesses. I can hear Chief Dinosaur complaining, 'This just turns us into bean counters. Besides, you can't assign ratings to emergency operations." Are we so unique that we cannot be accountable to a grading system? Admittedly, it is easier to figure profits and losses than it is to assign ratings to emergency operations, but using a report card can and should be done. Unfortunately, most fire departments don't evaluate IC's until something goes wrong, causing deaths, injuries, or a major property loss.

Post Incident Analysis Interviews

One way to find out what happened on the fireground is to ask the firefighters. One-on-one interviews with key personnel can be very illuminating. During one interview, a firefighter proudly said that he helped advance his engine's line, then helped the truck put up a ground ladder. Next, he searched a few bedrooms by himself, until he heard there was trouble opening the roof, so up he went. This really happened, but you would never learn this from the typical PIA. His officer would never tell you or write about it, because he probably doesn't know what really happened, and wouldn't talk about it if he did.

A ONE-ON-ONE FIREGROUND INTERVIEW

The Tale of the Tape

Audiotapes are pure gold when it comes to post incident analysis.

It is easy to record your fireground radio traffic. Often, the Dispatch Center records incidents. If this is not possible, taping can be done using a small battery-operated recorder placed in the Command Post. You will be amazed at the results. Audiotapes are pure gold when it comes to post incident analysis. The tape will show the flow of an incident, good and bad, and give a different perspective to the fireground. You will hear things that you didn't hear at the time of the incident, and the time sequence may seem faster or slower than you remember it.

**TAPE YOUR FIREGROUND
COMMUNICATIONS**

To use a tape in a post incident analysis, have the IC listen to the tape first and then let the companies or individuals listen to the tape on their own. Using this laid-back and low-key approach

286

is less threatening to personnel and generates interest and cooperation. You will find that the troops are very interested in listening to the tapes of their own firegrounds. After each individual or group has listened to the tape, the IC can sit down with them for an informal talk about how radio communications can be improved.

Reviewing audio tapes also gives personnel the opportunity to solve their own problems. The IC will often hear unsolicited comments such as, 'Next time I will give clearer reports."The best way to improve is to recognize your own errors and resolve to correct them. It's much better than the boss correcting them. The fireground tape allows the IC's to evaluate themselves. One chief officer who uses the tape evaluation system has this to say: 'I found t hat I was using some poor terminology. After listening, I would think of ways to improve for next time. This has improved me as an incident commander."

Fireground audiotapes are excellent self- improvement tools.

Let me tell you a little story about my taping experiences. After learning the value of fireground tapes, I decided that I would start taping my working incidents by using a small battery-powered tape recorder. It sounded like a good plan, but it didn't work very well. The problem was that at working incidents, I got busy and would forget to push the record button. After half a dozen failed attempts, I was feeling pretty stupid. I asked myself what was so hard about pushing a button when we had a worker? I finally solved the problem. I put the tape recorder on the seat of the chief's car where I have to pick it up to sit down. When I pick it up, I push the record button. I have done this on all calls. Now, if the call isn't a working fire (which it usually isn't), I just rewind the recorder and put it back on the seat. It's a simple system, but it works. Since I started using this system, I have recorded all my working fires.

The moral of the story goes beyond tape recording and into philosophy. The reason that the 'push it every time"system works is because it is part of my r egular routine. Trying to have two modes of operations, one for small incidents and another for big ones doesn't work, but approaching all incidents as if they are serious does work.

Making Post Incident Analysis Work

The best systems for recording or rating fireground operations won't do any good unless the post incident analysis process is accepted by your troops. The officers need to support the process and firefighters need to trust it. Some ideas for insuring PIA acceptance follow.

Critiques Should be Two-Way Streets

I had been reassigned to a different battalion where I was tightening up on fireground operations. I was monitoring the company officer's fireground performances very closely. I was fair but firm. We soon had a haz mat incident that was badly organized, something that was my fault. Later, one of the company officers asked me, 'How do you think this operat ion went chief?"I told him that I thought that it stunk, but that I was working on improving. He nodded knowingly and left. I had passed the test. I was holding myself to the same standards to which I was holding them. Although the operation was poor, I think that my credibility improved that day.

287

Use the PIA to Make Improvements

Regardless of which PIA methods are used, it is vital that the troops see the results showing the lessons learned or reinforced, and changes that were made. An example follows:

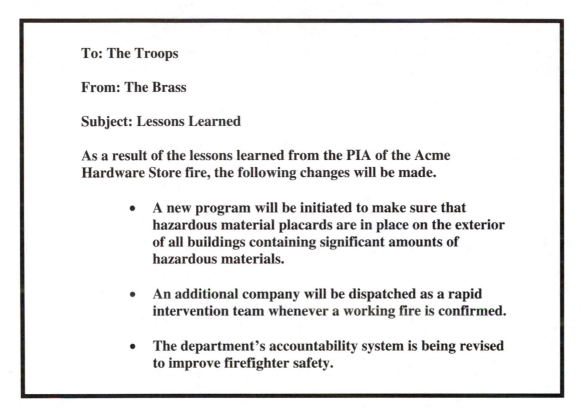

To: The Troops

From: The Brass

Subject: Lessons Learned

As a result of the lessons learned from the PIA of the Acme Hardware Store fire, the following changes will be made.

- **A new program will be initiated to make sure that hazardous material placards are in place on the exterior of all buildings containing significant amounts of hazardous materials.**

- **An additional company will be dispatched as a rapid intervention team whenever a working fire is confirmed.**

- **The department's accountability system is being revised to improve firefighter safety.**

This gives the troops confidence that the PIA is helpful, and that their fire department is progressive and interested in their well-being. Important information should also be shared with other departments.

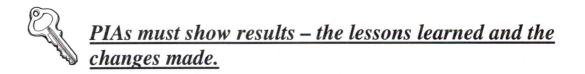

PIAs must show results – the lessons learned and the changes made.

Consider Changing Your Procedures

Recently an aerial ladder collapsed, carrying a firefighter who had been directing a stream from the top of the ladder. The firefighter narrowly avoided death when the ladder came crashing down. The investigation revealed the cause of the accident to be a structural failure of the ladder. As a result, the other ladder trucks in the fire department were tested to make sure that no other ladders were defective. This addressed the mechanical problems, but nobody ever looked at why the procedures allowed the firefighter to be on top of the ladder in the first place. There is a tendency to look more closely at mechanical defects than at poor procedures.

288

We Burned Our Firefighters, but It Went Great

Common sense tells you that if you have a simple garage fire and wind up with six firefighters in the burn unit, something's not right.

A volunteer department responded to a fire in a garage that was attached to the house. The first hose line went into the house to attack the garage fire, following department SOPs. The firefighters were inside a small vestibule leading into the garage when the fire flashed over, seriously burning six firefighters. They could not maneuver because they were jammed into a narrow space leading into the garage. The second line had gone to protect a distant exposure (a detached house), according to SOP. Their PIA stated that since all SOPs were followed, no errors were made. Common sense tells you that if you have a simple garage fire and wind up with six firefighters in the burn unit, something's not right.

The garage door was open, and it would have been easy to knock the fire down from the safety of the driveway. Once this was accomplished, the line should have gone into the house. The exposure next door was too far away to be an important consideration. Even if attacking the fire from the inside was a good idea (it wasn't), the crew was in a dangerous situation and was operating without the protection of a backup line. The backup line was critical to safety, but was wasted on the unlikely potential exposure. This fire fighting operation was seriously out of whack. The PIA should be used in this case to change things, not to justify the current practices within the department. **Use the PIA to look closely and critically at your SOP system.**

Culture Problems

An investigation was held following a fire that caused the death of one firefighter and serious injuries to other firefighters. Many of the officers and firefighters interviewed thought that the incident was a "routine" fire, until firefighters were injured. Following the fire, there was a general lack of understanding of what actually happened on the fire scene. Some were looking to blame one company, others did not understand that some of their actions contributed to the tragedy. When asked for their recommendations to prevent future tragedies, most thought that increased personnel would have changed the outcome of the incident.

Analysis of the fire showed that staffing appeared to have had little affect on the outcome. For example, there were serious ventilation problems because the windows were not promptly ventilated, yet the first alarm truck company had never attempted to ventilate the windows. Slow window ventilation can be blamed on reduced staffing, but not the failure to try to ventilate.

The fireground was very disorganized and chaotic. Company officers did not report that firefighters were missing, and once this became known it took too long to react, locate, and remove the injured firefighters. The investigation found that poor fireground behavior by both officers and firefighters contributed to the death and injuries, including the following:

- Not following established SOPs

- Not effectively communicating tactical information

- Not managing or coordinating the incident through the Incident Command System

- Members not staying together

- Freelancing

Additionally, the investigation revealed that many fire incidents in this department were poorly managed with companies, officers and firefighters operating outside of established systems and procedures. Fire incidents were usually brought under control quickly; however, when fire conditions escalated, poorly disciplined fire operations turn chaotic and dangerous.

> The investigation revealed that the problems at this fire were largely cultural. Fire operations had been lax for so long that screw-ups had become normal. The department had to change members' attitudes so that they would realize that SOPs and proper fireground behaviors are important for "routine" fires, not just "major" fires. To reduce the likelihood of future tragedies, all members had to realize that this tragic fire was the result of a badly flawed operation. The investigation stated that SOPs must be vigorously enforced and incidents regularly critiqued in order to improve operations. The department's culture could no longer permit any lack of professionalism or lack of responsibility for improper actions. A new culture was needed where firefighters follow the rules, and think and operate safely. The thinking, safety-oriented firefighter should be the role model, not the Kamikaze firefighter. Finally, the investigation concluded that changes must receive total support from the highest levels in the fire department in order to make all of the necessary improvements.

The lessons learned from this tragic incident can be applied to all fire departments. We all have some of these problems in our own fire departments. We need to recognize and correct our problems to prevent future tragedies.

Screwed-up PIAs

There was a fire in the cockloft of a typical McDonalds fast food restaurant. The fire had started in the grease duct and extended into the attic. Everyone was out of the restaurant upon the arrival of the fire department. They attacked the fire by sending engine companies inside to attack from below, while two truck companies were on the roof in an extensive roof-cutting operation.

The PIA for this fire was interesting. It stated that the fire was in the space below the roof and above the ceiling. There was no mention that the fire was in the truss area and that it was a classic collapse situation. The lessons learned were filled with the best ways to cut open the roof. The recommendations were basically better ways to do something dumb. The conclusion was that aggressive truck work on the roof and inside hose lines brought this fire under control. A job well done and a pat on the back for all.

The reality is that firefighters were exposed to great risks, both over and under the fire. There was no real gain. The plan should have been to set up a collapse zone and throw on lots of water. That would have been a low risk strategy, and the results would have been about the same.

Another PIA followed a serious fire that engulfed a townhouse. The report did a good job of telling what happened and what went wrong. It stated that the main problems were poor communications and poor coordination between the inside and outside hose streams. The problem was that the PIA never explained why communications and coordination had problems or what should be changed to prevent the problems from reoccurring.

Improving Written PIAs

One way to improve the PIA is to use a standard format to include the following sections:

- Description of the incident

- Strategy and tactics employed

- The potential risks and gains of the strategy used

- The lessons learned or reinforced

This standard format approach does not permit the person writing the report to tell you only what they want you to know. When using a standard format, all PIAs should be of similar quality and cover the same subjects, so that objective comparisons among reports can be made. This allows the PIA to do what it was designed for: improve operations, and not just be a fuzzy feel-good exercise.

SYSTEMS ANALYSIS

From time to time, the people in every organization need to figure out where they are, where they are going, and what needs improvement. Fire departments can become cluttered up like your desk. Every now and then, you have to sort through your desk, throw out the old stuff and organize what's left. If you don't do this, you can reach a critical clutter point where efficiency suffers. This is also true of fire departments. Sometimes things can get buried and don't get looked at or evaluated for years.

Many fire departments have some procedures that are outdated or just don't make sense. One department has a rule that when an aerial ladder is raised to the roof, the tip of the ladder should be extended one rung above the roof. It is a dumb rule because the aerial ladder should be a prominent escape beacon for firefighters on the roof. Sharp truck operators ignore it, but some operators still abide by the book way to do it. Another department washed the apparatus wheels after every call. Why? Because when they had steamers they had to clean the horse manure from the streets off the wheels, and the practice continued until several years ago.

In most departments, there is no easy way to change rules like this. Fire chiefs have never been big on suggestions boxes, so communications often go one way–from the top down. As a result, many departments operate with a hodge-podge of regulations, procedures and rules that have been added, piecemeal, over the years. Seldom will a department do a top-to-bottom analysis of their emergency operations to ensure that operations reflect modern times.

How do you analyze your fire departments operations? One way is to get your key people together (including some front line troops) annually and have an open discussion of your emergency operations. Be sure to include people from other fire departments, people who can add to your perspective. Chief Dinosaur roars, 'We have plenty of good people! We don't need anyone from other departments." Chief D is right, h e does have plenty of good people. The problem is that, within the same department, they all tend to think alike. A knowledgeable outsider comes in without preconceived ideas, and can often make valuable suggestions.

Look at other fire departments and how they operate. How do your operations vary from those of your neighbors and similar sized departments? You will find that they do some things better than you do. Look at NFPA recommendations and legal requirements to see if your department is in compliance. Come up with a plan for improvement, and do it.

SUMMARY

Analyze past incidents to improve your operations. Firefighters often die as a result of fireground problems that have surfaced time and again, but were never addressed.

Three major factors contributing to firefighters deaths are: communications problems, lack of accountability, and poor strategy and tactics.

It is important to evaluate an incident quickly after it occurs. Create a relaxed non-threatening atmosphere. Focus on safety because everyone considers safety to be a high priority item.

Use a standard format for written post incident analysis reports.

The ICS and SOPs can be used to compare what happened against what was supposed to happen. If something was not SOP but worked well, or if the SOP operation was poor, review the SOP to see if it needs to be changed or updated.

Tape record and critique your fireground radio traffic to improve your operations.

The principal post incident analysis methods are: the informal method, the formal method, written evaluations and interview methods.

When using comparison analysis (a written evaluation method), each officer completes a written report on what they did, why they did it, problems found, and the results. When all forms are completed, they are compared to show what worked well and what didn't.

By using the IC report card (a written evaluation method), all incidents are evaluated using numerical ratings. Incidents can then be compared to show patterns of strengths and weaknesses in your operation.

The post incident analysis gives the troops confidence that the fire department is progressive, improving, and interested in their well-being.

Conduct a top-to-bottom review of your department's emergency operations annually.

22

CONCLUSION

The Future

The fire service will see many changes in the future. Firefighters will have thermal imaging as an integral part of the facemask. Remaining air time will be displayed in a corner of the facepiece, and detectors will warn of dangerous heat buildup or hazardous materials. When the firefighters turn to exit, they will be guided to safety by visual flashes and audio signals showing the way out.

Command will have the building's floor plan and pre-plans on computer screens at the command post. A tracking system will show the location of each firefighter inside the building. Portable radios will be carried by all firefighters and messages will be sent without fumbling for buttons. Radios will be reliable at all fires and emergencies. Portable radio channels can be changed from the command post as incidents escalate. Command charts will be like computer games, so that the IC can move and track sectors and companies with a joy stick. There will be more dependency on mutual aid and outside resources, and we will be better prepared for large-scale fires, disasters and terrorism.

These are some of the things on our wish list. Many of these things will happen, and many unforeseen changes will happen. The one constant in the future is the role of our people. In the future as now, we will always need to be guided by common sense. I would like to leave you with a collection of rules for you to ponder.

Dumb Rules

The following is a collection of dumb rules from across the country, presented for your amusement and education. If you agree with more than two of them, it's time for you to take your Firefighter 1 test again.

- All fires are different, so SOPs don't work very well.

- After twenty minutes, collapse can be anticipated.

- Cut the vent hole right over the fire.

293

- The faster you drive, the more lives you save.

- Building walls always fall out one-third their height.

- Park the ladder truck out of the way.

- It can't happen here.

- If you have anyone left over, make that person the Safety Officer.

- Speed extinguishment by using ladder truck firefighters on hose lines.

- Start using ICS when things start to get out of hand.

- Opening walls and ceilings causes unnecessary damage.

- Reduce damage; don't break glass.

- If fire vents from the vent hole, try to make the hole bigger.

- First in, last out.

- Haz mat control zones are for the civilians.

- Set up all the ICS first; then worry about the fire.

- No extinguishment, no rehab.

- Fight fires with the fewest resources that you can.

- The less you say on the radio the better.

- The more you say on the radio the better.

- Mutual aid is a necessary evil.

- First find the fire; then get prepared.

- Lay a supply line only when you see smoke and fire.

- Accountability checks can wait until Command gets around to it.

- Five hundred gallons will be enough.

- Always hang in there until the pack alarm starts to ring, and then stay a little longer.

- Send the second line to the exposures; exposures are more important than the safety of your attack crew.

- The last firefighter wearing a facepiece is a wussy.

- My officer is responsible for my safety.

- You don't need two trucks at a working fire.

- Changing fireground channels in the middle of a fire is easy.

- The best way to test the air is with your nose.

- Command should call for help one company at a time.

- Command should wait as long as possible before changing into a defensive operation: those extra minutes could do the trick.

- The first thing to do at water rescues is jump into the water.

- Don't use accountability unless the fire is a second alarm or better.

- Look for bubbling tar, melted snow, or dry spots on a wet roof. You then know where to stand to open the roof over the fire.

Smart Rules to Live By

The following are a few words of wisdom. They come from various fire departments as well as from this book. If you are already doing all of these things, you probably should have written this book instead of reading it.

- The first commandment for a fireground commander is: Take care of your troops.

- The best guidelines are usually simple; don't get too fancy.

- SOPs ensure that the correct procedures will happen with few orders given, use them at every fire.

- Command should use a stationary command post, and use command charts and check-off systems.

- Engine companies rescue people by putting out the fire.

- Make sure that truck work gets done.

- Get ahead of the fire, don't chase it.

- Always use the minimum number of personnel necessary in any hazard zone.

- Every incident is an opportunity to learn.

- Operations at small fires set the stage for future big fires.

- At most fires, there are only a few key points that are really important. The trick is to focus on those critical to the incident.

- When the engine crew is inside attacking, their safety must be your first priority.

- A full staging area can be the IC's ace in the hole.

- Call for help based on what could happen.

- Common sense will help you with all incidents.

- Be cool. Losing your head can be dangerous.

- Check your equipment regularly and faithfully.

- There is no such thing as a routine call.

- Complacency kills.

- Use a Mayday call for firefighters in trouble.

- Above all, look out for the safety of the firefighters.

- Training must be practical and relate to the real world.

- After incidents talk it up: what happened and how things could be improved.

- Dispatch response levels should be increased when there is a strong possibility of a working incident. The greater the problem, the more you send.

- We should prepare for the worst circumstances, not the easy ones.

- There are no small fire departments.

- Use vests and ICS cheat sheets.

- Good pre-plans make firegrounds safer.

- What we do on everyday incidents becomes a habit that is almost impossible to change in the heat of battle.

- Trucks should be positioned for aerial ladder use at each alarm.

- Above all, look out for the safety of the firefighters.

- The first arriving engine should establish its own water supply, if possible by laying a supply line from a hydrant.

- Never be lazy on the fireground.

- Don't pass by fire. Knock it down thoroughly before you advance.

- Never let fire get behind you, and don't get ahead of your line.

- Think master streams at any doubtful operation.

- Don't drive fire back into the roof.

- Master streams must never be used with an interior attack.

- A Rehab Sector should be established at every working incident.

- One engine company in the right place at the right time is worth ten engines later on.

- Backup lines are vital for safety and efficiency.

- Attack methods are not carved in stone. If there are problems, change the plan.

- Use SOPs at every fire.

- A good fireground designation system allows everyone on the fireground to quickly identify any location at a scene.

- Avoid going down the stairs at working basement fires.

- Avoid mixing hand lines with master streams.

- A basic principle is to attack basement fires from the same level, not from above.

- Avoid using elevators if possible.

- Regularly communicate on all alarms so everyone becomes accustomed to working and talking with each other. Then, communications will run smoothly at workers.

- Don't play musical chairs with radio channels.

- Record your fireground radio traffic for post incident analysis.

- Safety must be part of your everyday operations.

- Beware of nonresidential structure fires, they are very dangerous.

- Above all, look out for the safety of the firefighters.

- Focus on the software in the drivers' heads, not the color of the apparatus.

- Use controlled aggression, and weigh the potential gain against the loss.

- Develop and practice lost firefighter SOPs.

- No one operates alone in the hazard area.

- The crew that sticks together, gets out together.

- Your accountability system needs to be simple, accurate, and used on every call.

- Accountability is everyone's job.

- Check the basement and the floors below the reported fire floor to avoid ugly surprises.

- Rapid intervention teams save lives.

- Smart firefighters don't breathe poisons.

- The roof team is often in an excellent position to report overall conditions.

- Avoid the tendency to relax and let your guard down as time goes on.

- Fire fighting inside poorly ventilated structures is dangerous, and glass is cheap.

- Haz mat control zones should be set up early and at the proper distances.

- The fastest and safest way to ventilate windows is to use ground ladders to break out the windows.

- Above all, look out for the safety of the firefighters.

- If you're thinking about looking for signs of collapse, you probably shouldn't be in the building.

- Conduct a safety check of structures before beginning to overhaul.

- Protect exposures on all sides, inside and out.

- If the place is going to burn down, do it with class.

- Use fast, safe roof operations.

- When the outcome is in doubt, call for help.

- Call for help quickly and in force, and avoid dribbling in help.

- The IC needs aides at major incidents.

- Beware of fire situations where collapse is likely.

- Use checklists.

- Don't play musical chairs with chiefs. Keep the former IC at the command post to begin building a command team.

- In haz mat, sometimes the best decision is to take no action, or a limited action.

- First in, first out.

- Have a plan and a backup plan.

- The IC should talk to sectors not individual companies.

- Sector all multi-company incidents.

- Maintain control through staging.

- Assume nothing. Do the basics properly on each run.

- Use ICS on every call.

- Vent early and vent often.

- Use a Safety Officer at all working incidents.

- Putting out the fire solves most of the problems.

- When in doubt, play it safe.

- Above all, look out for the safety of the firefighters.

I hope that you have learned a few things from looking over this book. If you walk away with only one message, it should be: **Always do it the right way. It isn't always necessary, but it will ensure that you will be doing it the right way when it counts**.

INDEX

ABOUT THE AUTHOR

Robert C. Bingham, a 31-year veteran of the fire service, is a fire training specialist and consults with fire departments investigating firefighter fatalities. He formerly was a deputy chief in the District of Columbia Fire Department, where he served as a command officer for 15 years and implemented standard operating procedures for structure fires and the Incident Command System in Washington D. C. He has authored many articles on fire fighting procedures. He has taught at the National Fire Academy and as an instructor and course developer for the University of the District of Columbia Fire Science program. He is a graduate of the University of Maryland, University College, at College Park Maryland. He lives in Vienna, Virginia with his wife Joann.

Bob Bingham

The author welcomes your comments and suggestions. He can be reached at binghamrj@hotmail.com

Order this book today!
At: streetsmartfire.com
Or Call Toll Free: 1-888-825-9550
Fax Your Order: 703-281-2994

STREET SMART FIREFIGHTING	QTY.	PRICE	TOTAL
(Virginia Residents add 5% sales tax)		$29.95	
SHIPPING AND HANDLING		*FREE*	
		TOTAL	

Send check to:

Valley Press
Post Office Box 2044
Vienna, VA 22183

Enclosed is my check/money order____   <u>Credit card orders</u>

Print Name   Visa ____ Master card ____

_____   **Name on credit card**

Address

_____   _____

City   **State**   **Zip**   **Credit card # (include expiration date)**

_____   _____

Thank you for your order!